Introduction to Quantum Mechanics

The Manchester Physics Series

General Editors
D. J. SANDIFORD: F. MANDL: A. C. PHILLIPS

Department of Physics and Astronomy,
University of Manchester

Properties of Matter:	B. H. Flowers and E. Mendoza
Statistical Physics: *Second Edition*	F. Mandl
Electromagnetism: *Second Edition*	I. S. Grant and W. R. Phillips
Statistics:	R. J. Barlow
Solid State Physics: *Second Edition*	J. R. Hook and H. E. Hall
Quantum Mechanics:	F. Mandl
Particle Physics: *Second Edition*	B. R. Martin and G. Shaw
The Physics of Stars: *Second Edition*	A. C. Phillips
Computing for Scientists:	R. J. Barlow and A. R. Barnett
Nuclear Physics:	J. S. Lilley
Introduction to Quantum Mechanics:	A. C. Phillips

INTRODUCTION TO QUANTUM MECHANICS

A. C. Phillips
Department of Physics and Astronomy
University of Manchester

WILEY

Other Wiley Editorial Offices

John Wiley & Sons Inc., 111 River Street, Hoboken, NJ 07030, USA

Jossey-Bass, 989 Market Street, San Francisco, CA 94103-1741, USA

Wiley-VCH Verlag GmbH, Boschstr. 12, D-69469 Weinheim, Germany

John Wiley & Sons Australia Ltd, 33 Park Road, Milton, Queensland 4064, Australia

John Wiley & Sons (Asia) Pte Ltd, 2 Clementi Loop #02-01, Jin Xing Distripark, Singapore 129809

John Wiley & Sons Canada Ltd, 22 Worcester Road, Etobicoke, Ontario, Canada M9W 1L1

British Library Cataloguing in Publication Data

A catalogue record for this book is available from the British Library

ISBN 13: 978-0-470-85323-8 (H/B)   ISBN 13: 978-0-470-85324-5 (P/B)

Typeset by Kolam Information Services Pvt. Ltd. Pondicherry, India
Printed and bound in Great Britain by Antony Rowe Ltd, Chippenham, Wiltshire.

To my sons:

Joseph
Michael
Patrick
Peter

Contents

Foreword

Sadly, Tony Phillips, a good friend and colleague for more than thirty years, died on 27th November 2002. Over the years, we discussed most topics under the sun. The originality and clarity of his thoughts and the ethical basis of his judgements always made this a refreshing exercise. When discussing physics, quantum mechanics was a recurring theme which gained prominence after his decision to write this book. He completed the manuscript three months before his death and asked me to take care of the proofreading and the Index. A labour of love. I knew what Tony wanted—and what he did not want. Except for corrections, no changes have been made.

Tony was an outstanding teacher who could talk with students of all abilities. He had a deep knowledge of physics and was able to explain subtle ideas in a simple and delightful style. Who else would refer to the end-point of nuclear fusion in the sun as sunshine? Students appreciated him for these qualities, his straightforwardness and his genuine concern for them. This book is a fitting memorial to him.

Franz Mandl
December 2002

Editors' preface to the Manchester Physics Series

The Manchester Physics Series is a series of textbooks at first degree level. It grew out of our experience at the Department of Physics and Astronomy at Manchester University, widely shared elsewhere, that many textbooks contain much more material than can be accommodated in a typical undergraduate course; and that this material is only rarely so arranged as to allow the definition of a short self-contained course. In planning these books we have had two objectives. One was to produce short books: so that lecturers should find them attractive for undergraduate courses; so that students should not be frightened off by their encyclopaedic size or price. To achieve this, we have been very selective in the choice of topics, with the emphasis on the basic physics together with some instructive, stimulating and useful applications. Our second objective was to produce books which allow courses of different lengths and difficulty to be selected with emphasis on different applications. To achieve such flexibility we have encouraged authors to use flow diagrams showing the logical connections between different chapters and to put some topics in starred sections. These cover more advanced and alternative material which is not required for the understanding of latter parts of each volume.

Although these books were conceived as a series, each of them is self-contained and can be used independently of the others. Several of them are suitable for wider use in other sciences. Each Author's Preface gives details about the level, prerequisites, etc., of that volume.

The Manchester Physics Series has been very successful with total sales of more than a quarter of a million copies. We are extremely grateful to the many students and colleagues, at Manchester and elsewhere, for helpful criticisms and stimulating comments. Our particular thanks go to the authors for all the work they have done, for the many new ideas they have contributed, and for discussing patiently, and often accepting, the suggestions of the editors.

Finally we would like to thank our publishers, John Wiley & Sons, Ltd, for their enthusiastic and continued commitment to the Manchester Physics Series.

D. J. Sandiford
F. Mandl
A. C. Phillips
February 1997

Author's preface

There are many good advanced books on quantum mechanics but there is a distinct lack of books which attempt to give a serious introduction at a level suitable for undergraduates who have a tentative understanding of mathematics, probability and classical physics.

This book introduces the most important aspects of quantum mechanics in the simplest way possible, but challenging aspects which are essential for a meaningful understanding have not been evaded. It is an introduction to quantum mechanics which

- motivates the fundamental postulates of quantum mechanics by considering the weird behaviour of quantum particles
- reviews relevant concepts in classical physics before corresponding concepts are developed in quantum mechanics
- presents mathematical arguments in their simplest form
- provides an understanding of the power and elegance of quantum mechanics that will make more advanced texts accessible.

Chapter 1 provides a qualitative description of the remarkable properties of quantum particles, and these properties are used as the guidelines for a theory of quantum mechanics which is developed in Chapters 2, 3 and 4. Insight into this theory is gained by considering square wells and barriers in Chapter 5 and the harmonic oscillator in Chapter 6. Many of the concepts used in the first six chapters are clarified and developed in Chapter 7. Angular momentum in quantum mechanics is introduced in Chapter 8, but because angular momentum is a demanding topic, this chapter focusses on the ideas that are needed for an understanding of the hydrogen atom in Chapter 9, identical particles in Chapter 10 and many-electron atoms in Chapter 11. Chapter 10 explains why identical particles are described by entangled quantum states and how this entanglement for electrons leads to the Pauli exclusion principle.

Chapters 7 and 10 may be omitted without significant loss of continuity. They deal with concepts which are not needed elsewhere in the book.

xvi   Author's preface

I would like to express my thanks to students and colleagues at the University of Manchester. Daniel Guise helpfully calculated the energy levels in a screened Coulomb potential. Thomas York used his impressive computing skills to provide representations of the position probabilities for particles with different orbital angular momentum. Sean Freeman read an early version of the first six chapters and provided suggestions and encouragement. Finally, I would like to thank Franz Mandl for reading an early version of the book and for making forcefully intelligent suggestions for improvement.

A. C. Phillips
August 2002

1

Planck's constant in action

Classical physics is dominated by two fundamental concepts. The first is the concept of a particle, a discrete entity with definite position and momentum which moves in accordance with Newton's laws of motion. The second is the concept of an electromagnetic wave, an extended physical entity with a presence at every point in space that is provided by electric and magnetic fields which change in accordance with Maxwell's laws of electromagnetism. The classical world picture is neat and tidy: the laws of particle motion account for the material world around us and the laws of electromagnetic fields account for the light waves which illuminate this world.

This classical picture began to crumble in 1900 when Max Planck published a theory of black-body radiation; i.e. a theory of thermal radiation in equilibrium with a perfectly absorbing body. Planck provided an explanation of the observed properties of black-body radiation by assuming that atoms emit and absorb discrete quanta of radiation with energy $\epsilon = h\nu$, where ν is the frequency of the radiation and h is a fundamental constant of nature with value

$$h = 6.626 \times 10^{-34} \, \text{J s}.$$

This constant is now called Planck's constant.

In this chapter we shall see that Planck's constant has a strange role of linking wave-like and particle-like properties. In so doing it reveals that physics cannot be based on two distinct, unrelated concepts, the concept of a particle and the concept of a wave. These classical concepts, it seems, are at best approximate descriptions of reality.

1.1 PHOTONS

Photons are particle-like quanta of electromagnetic radiation. They travel at the speed of light c with momentum p and energy ϵ given by

$$p = \frac{h}{\lambda} \quad \text{and} \quad \epsilon = \frac{hc}{\lambda},$$ (1.1)

where λ is the wavelength of the electromagnetic radiation. In comparison with macroscopic standards, the momentum and energy of a photon are tiny. For example, the momentum and energy of a visible photon with wavelength $\lambda = 663$ nm are

$$p = 10^{-27} \text{ J s} \quad \text{and} \quad \epsilon = 3 \times 10^{-19} \text{ J}.$$

We note that an electronvolt, $1 \text{ eV} = 1.602 \times 10^{-19}$ J, is a useful unit for the energy of a photon: visible photons have energies of the order of an eV and X-ray photons have energies of the order of 10 keV.

The evidence for the existence of photons emerged during the early years of the twentieth century. In 1923 the evidence became compelling when A. H. Compton showed that the wavelength of an X-ray increases when it is scattered by an atomic electron. This effect, which is now called *the Compton effect*, can be understood by assuming that the scattering process is a photon–electron collision in which energy and momentum are conserved. As illustrated in Fig. 1.1, the incident photon transfers momentum to a stationary electron so that the scattered photon has a lower momentum and hence a longer wavelength. In fact, when the photon is scattered through an angle θ by a stationary electron of mass m_e, the increase in wavelength is given by

$$\Delta\lambda = \frac{h}{m_e c}(1 - \cos\theta).$$ (1.2)

We note that the magnitude of this increase in wavelength is set by

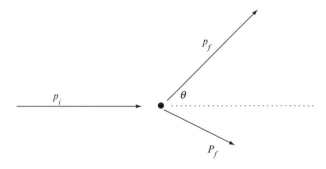

Fig. 1.1   A photon–electron collision in which a photon is scattered by a stationary electron through an angle θ. Because the electron recoils with momentum P_f, the magnitude of the photon momentum decreases from p_i to p_f and the photon wavelength increases.

$$\frac{h}{m_e c} = 2.43 \times 10^{-12}\,\text{m},$$

a fundamental length called the *Compton wavelength* of the electron.

The concept of a photon provides a natural explanation of the Compton effect and of other particle-like electromagnetic phenomena such as the photo-electric effect. However, it is not clear how the photon can account for the wave-like properties of electromagnetic radiation. We shall illustrate this difficulty by considering the two-slit interference experiment which was first used by Thomas Young in 1801 to measure the wavelength of light.

The essential elements of a two-slit interference are shown in Fig. 1.2. When electromagnetic radiation passes through the two slits it forms a pattern of interference fringes on a screen. These fringes arise because wave-like disturbances from each slit interfere constructively or destructively when they arrive at the screen. But a close examination of the interference pattern reveals that it is the result of innumerable photons which arrive at different points on the screen, as illustrated in Fig. 1.3. In fact, when the intensity of the light is very low, the interference pattern builds up slowly as photons arrive, one by one, at random points on the screen after seemingly passing through both slits in a wave-like way. These photons are not behaving like classical particles with well-defined trajectories. Instead, when presented with two possible trajectories, one for each slit, they seem to pass along both trajectories, arrive at a random point on the screen and build up an interference pattern.

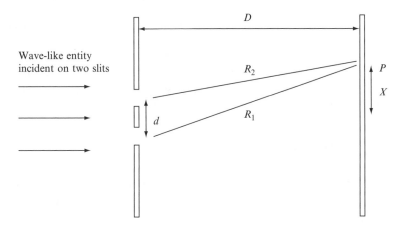

Fig. 1.2 A schematic illustration of a two-slit interference experiment consisting of two slits with separation d and an observation screen at distance D. Equally spaced bright and dark fringes are observed when wave-like disturbances from the two slits interfere constructively and destructively on the screen. Constructive interference occurs at the point P, at a distance x from the centre of the screen, when the path difference $R_1 - R_2$ is an integer number of wavelengths. This path difference is equal to xd/D if $d \ll D$.

Pattern formed by 100 quantum particles

Pattern formed by 1000 quantum particles

Pattern formed by 10 000 quantum particles

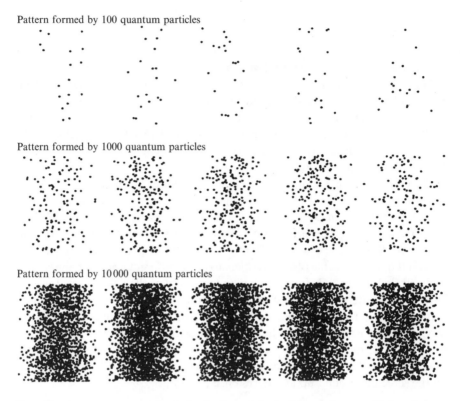

Fig. 1.3 A computer generated simulation of the build-up of a two-slit interference pattern. Each dot records the detection of a quantum particle on a screen positioned behind two slits. Patterns formed by 100, 1000 and 10 000 quantum particles are illustrated.

At first sight the particle-like and wave-like properties of the photon are strange. But they are not peculiar. We shall soon see that electrons, neutrons, atoms and molecules also behave in this strange way.

1.2 DE BROGLIE WAVES

The possibility that particles of matter like electrons could be both particle-like and wave-like was first proposed by Louis de Broglie in 1923. Specifically he proposed that a particle of matter with momentum p could act as a wave with wavelength

$$\lambda = \frac{h}{p}. \tag{1.3}$$

This wavelength is now called the *de Broglie wavelength*.

It is often useful to write the de Broglie wavelength in terms of the energy of the particle. The general relation between the relativistic energy ϵ and the momentum p of a particle of mass m is

$$\epsilon^2 - p^2 c^2 = m^2 c^4. \tag{1.4}$$

This implies that the de Broglie wavelength of a particle with relativistic energy ϵ is given by

$$\lambda = \frac{hc}{\sqrt{(\epsilon - mc^2)(\epsilon + mc^2)}}. \tag{1.5}$$

When the particle is ultra-relativistic we can neglect mass energy mc^2 and obtain

$$\lambda = \frac{hc}{\epsilon}, \tag{1.6}$$

an expression which agrees with the relation between energy and wavelength for a photon given in Eq. (1.1). When the particle is non-relativistic, we can set

$$\epsilon = mc^2 + E,$$

where $E = p^2/2m$ is the kinetic energy of a non-relativistic particle, and obtain

$$\lambda = \frac{h}{\sqrt{2mE}}. \tag{1.7}$$

In practice, the de Broglie wavelength of a particle of matter is small and difficult to measure. However, we see from Eq. (1.7) that particles of lower mass have longer wavelengths, which implies that the wave properties of the lightest particle of matter, the electron, should be the easiest to detect. The wavelength of a non-relativistic electron is obtained by substituting $m = m_e = 9.109 \times 10^{-31}$ kg into Eq. (1.7). If we express the kinetic energy E in electron volts, we obtain

$$\lambda = \sqrt{\frac{1.5}{E}} \text{ nm}. \tag{1.8}$$

From this equation we immediately see that an electron with energy of 1.5 eV has a wavelength of 1 nm and that an electron with energy of 15 keV has a wavelength of 0.01 nm.

Because these wavelengths are comparable with the distances between atoms in crystalline solids, electrons with energies in the eV to keV range are diffracted by crystal lattices. Indeed, the first experiments to demonstrate the wave properties of electrons were crystal diffraction experiments by C. J. Davisson and L. H. Germer and by G. P. Thomson in 1927. Davisson's experiment involved electrons with energy around 54 eV and wavelength 0.17 nm which were diffracted by the regular array of atoms on the surface of a crystal of nickel. In Thomson's experiment, electrons with energy around 40 keV and wavelength 0.006 nm were passed through a polycrystalline target and diffracted by randomly orientated microcrystals. These experiments showed beyond doubt that electrons can behave like waves with a wavelength given by the de Broglie relation Eq. (1.3).

Since 1927, many experiments have shown that protons, neutrons, atoms and molecules also have wave-like properties. However, the conceptual implications of these properties are best explored by reconsidering the two-slit interference experiment illustrated in Fig. 1.2. We recall that a photon passing through two slits gives rise to wave-like disturbances which interfere constructively and destructively when the photon is detected on a screen positioned behind the slits. Particles of matter behave in a similar way. A particle of matter, like a photon, gives rise to wave-like disturbances which interfere constructively and destructively when the particle is detected on a screen. As more and more particles pass through the slits, an interference pattern builds up on the observation screen. This remarkable behaviour is illustrated in Fig. 1.3.

Interference patterns formed by a variety of particles passing through two slits have been observed experimentally. For example, two-slit interference patterns formed by electrons have been observed by A. Tonomura, J. Endo, T. Matsuda, T. Kawasaki and H. Exawa (*American Journal of Physics*, vol. 57, p. 117 (1989)). They also demonstrated that a pattern still emerges even when the source is so weak that only one electron is in transit at any one time, confirming that each electron seems to pass through both slits in a wave-like way before detection at a random point on the observation screen. Two-slit interference experiments have been carried out using neutrons by R. Gähler and A. Zeilinger (*American Journal of Physics*, vol. 59, p. 316 (1991)), and using atoms by O. Carnal and J. Mlynek (*Physical Review Letters*, vol. 66, p. 2689 (1991)). Even molecules as complicated as C_{60} molecules have been observed to exhibit similar interference effects as seen by M. Arndt *et al.* (*Nature*, vol. 401, p. 680 (1999)).

These experiments demonstrate that particles of matter, like photons, are not classical particles with well-defined trajectories. Instead, when presented with two possible trajectories, one for each slit, they seem to pass along both trajectories in a wave-like way, arrive at a random point on the screen and build up an interference pattern. In all cases the pattern consists of fringes with a spacing of $\lambda D/d$, where d is the slit separation, D is the screen distance and λ is the de Broglie wavelength given by Eq. (1.3).

Physicists have continued to use the ambiguous word *particle* to describe these remarkable microscopic objects. We shall live with this ambiguity, but we shall occasionally use the term *quantum particle* to remind the reader that the object under consideration has particle and wave-like properties. We have used this term in Fig. 1.3 because this figure provides a compelling illustration of particle and wave-like properties. Finally, we emphasize the role of Planck's constant in linking the particle and wave-like properties of a quantum particle. If Planck's constant were zero, all de Broglie wavelengths would be zero and particles of matter would only exhibit classical, particle-like properties.

1.3   ATOMS

It is well known that atoms can exist in states with discrete or quantized energy. For example, the energy levels for the hydrogen atom, consisting of an electron and a proton, are shown in Fig. 1.4. Later in this book we shall show that bound states of an electron and a proton have quantized energies given by

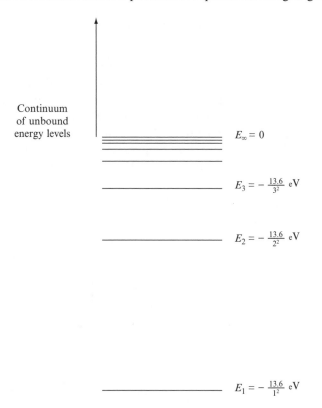

Fig. 1.4   A simplified energy level diagram for the hydrogen atom. To a good approximation the bound states have quantized energies given by $E_n = -13.6/n^2$ eV where n, the principal quantum number, can equal $1, 2, 3, \ldots$. When the excitation energy is above 13.6 eV, the atom is ionized and its energy can, in principle, take on any value in the continuum between $E = 0$ and $E = \infty$.

$$E_n = -\frac{13.6}{n^2} \text{ eV}, \tag{1.9}$$

where n is a number, called the *principal quantum number*, which can take on an infinite number of the values, $n = 1, 2, 3, \ldots$. The ground state of the hydrogen atom has $n = 1$, a first excited state has $n = 2$ and so on. When the excitation energy is above 13.6 eV, the electron is no longer bound to the proton; the atom is ionized and its energy can, in principle, take on any value in the continuum between $E = 0$ and $E = \infty$.

The existence of quantized atomic energy levels is demonstrated by the observation of electromagnetic spectra with sharp spectral lines that arise when an atom makes a transition between two quantized energy levels. For example, a transition between hydrogen-atom states with n_i and n_f leads to a spectral line with a wavelength λ given by

$$\frac{hc}{\lambda} = |E_{n_i} - E_{n_f}|.$$

Some of the spectral lines of atomic hydrogen are illustrated in Fig. 1.5.

Quantized energy levels of atoms may also be revealed by scattering processes. For example, when an electron passes through mercury vapour it has a high probability of losing energy when its energy exceeds 4.2 eV, which is the quantized energy difference between the ground and first excited state of a mercury atom. Moreover, when this happens the excited mercury atoms subsequently emit photons with energy $\epsilon = 4.2$ eV and wavelength

$$\lambda = \frac{hc}{\epsilon} = 254 \text{ nm}.$$

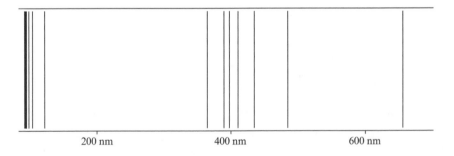

| 200 nm | 400 nm | 600 nm |

Fig. 1.5 Spectral lines of atomic hydrogen. The series of lines in the visible part of the electromagnetic spectrum, called the Balmer series, arises from transitions between states with principal quantum number $n = 3, 4, 5, \ldots$ and a state with $n = 2$. The series of lines in the ultraviolet, called the Lyman series, arises from transitions between states with principal quantum number $n = 2, 3, \ldots$ and the ground state with $n = 1$.

But quantized energy levels are not the most amazing property of atoms. Atoms are surprisingly resilient: in most situations they are unaffected when they collide with neighbouring atoms, but if they are excited by such encounters they quickly return to their original pristine condition. In addition, atoms of the same chemical element are identical: somehow the atomic number Z, the number of electrons in the atom, fixes a specific identity which is common to all atoms with this number of electrons. Finally, there is a wide variation in chemical properties, but there is a surprisingly small variation in size; for example, an atom of mercury with 80 electrons is only three times bigger than a hydrogen atom with one electron.

These remarkable properties show that atoms are not mini solar systems in which particle-like electrons trace well-defined, classical orbits around a nucleus. Such an atom would be unstable because the orbiting electrons would radiate electromagnetic energy and fall into the nucleus. Even in the absence of electromagnetic radiation, the pattern of orbits in such an atom would change whenever the atom collided with another atom. Thus, this classical picture cannot explain why atoms are stable, why atoms of the same chemical element are always identical or why atoms have a surprisingly small variation in size.

In fact, atoms can only be understood by focussing on the wave-like properties of atomic electrons. To some extent atoms behave like musical instruments. When a violin string vibrates with definite frequency, it forms a standing wave pattern of specific shape. When wave-like electrons, with definite energy, are confined inside an atom, they form a wave pattern of specific shape. An atom is resilient because, when left alone, it assumes the shape of the electron wave pattern of lowest energy, and when the atom is in this state of lowest energy there is no tendency for the electrons to radiate energy and fall into the nucleus. However, atomic electrons can be excited and assume the shapes of wave patterns of higher quantized energy.

One of the most surprising characteristics of electron waves in an atom is that they are *entangled* so that it is not possible to tell which electron is which. As a result, the possible electron wave patterns are limited to those that are compatible with a principle called the *Pauli exclusion principle*. These patterns, for an atom with an atomic number Z, uniquely determine the chemical properties of all atoms with this atomic number.

All these ideas will be considered in more detail in subsequent chapters, but at this stage we can show that the wave nature of atomic electrons provides a natural explanation for the typical size of atoms. Because the de Broglie wavelength of an electron depends upon the magnitude of Planck's constant h and the electron mass m_e, the size of an atom consisting of wave-like electrons also depends upon h and m_e. We also expect a dependence on the strength of the force which binds an electron to a nucleus; this is proportional to $e^2/4\pi\epsilon_0$, where e is the magnitude of the charge on an electron and on a proton. Thus, the order of magnitude of the size of atoms is expected to be a function of

$e^2/4\pi\epsilon_0$, m_e and h (or $\hbar = h/2\pi$). In fact, the natural unit of length for atomic size is the Bohr radius which is given by[1]

$$a_0 = \left[\frac{4\pi\epsilon_0}{e^2}\right]\frac{\hbar^2}{m_e} = 0.529 \times 10^{-10}\,\text{m}. \qquad (1.10)$$

Given this natural length, we can write down a natural unit for atomic binding energies. This is called the *Rydberg energy* and it is given by

$$E_R = \frac{e^2}{8\pi\epsilon_0 a_0} = 13.6\,\text{eV}. \qquad (1.11)$$

We note, that the binding energy of a hydrogen-atom state with principal quantum number n is E_R/n^2.

The Bohr radius was introduced by Niels Bohr in 1913 in a paper which presented a very successful model of the atom. Even though the Bohr model is an out-dated mixture of classical physics and ad-hoc postulates, the central idea of the model is still relevant. This idea is that Planck's constant has a key role in the mechanics of atomic electrons. Bohr expressed the idea in the following way:

> The result of the discussion of these questions seems to be the general acknowledgment of the inadequacy of the classical electrodynamics in describing the behaviour of systems of atomic size. Whatever alteration in the laws of motion of electrons may be, it seems necessary to introduce in the laws in question a quantity foreign to the classical electrodynamics; i.e., Planck's constant, or as it is often called, the elementary quantum of action. By introduction of this quantity the question of the stable configuration of the electrons in atoms is essentially changed, as this constant is of such dimensions and magnitude that it, together with the mass and the charge of the particles, can determine a length of the order of the magnitude required.

Ten years after this was written, it was realised that Planck's constant has a role in atoms because it links the particle-like and wave-like properties of atomic electrons.

1.4   MEASUREMENT

In classical physics, the act of measurement need not affect the object under observation because the disturbance associated with the measurement can be

---

[1] In this dimensional analysis argument we have not included a dependence on the velocity of light c because electrons in atoms are, to a good approximation, non-relativistic.

made arbitrarily small. Accordingly, the properties of a classical object can be specified with precision and without reference to the process of measurement. This is not the case in quantum physics. Here measurement plays an active and disturbing role. Because of this, quantum particles are best described within the context of the possible outcomes of measurements. We shall illustrate the role of measurement in quantum mechanics by introducing the Heisenberg uncertainty principle and then use this principle to show how measurement provides a framework for describing particle-like and wave-like quantum particles.

The uncertainty principle

We shall introduce the uncertainty principle for the position and momentum of a particle by considering a famous thought experiment due to Werner Heisenberg in which the position of a particle is measured using a microscope. The particle is illuminated and the scattered light is collected by the lens of a microscope as shown in Fig. 1.6.

Because of the wave-like properties of light, the microscope has a finite spatial resolving power. This means that the position of the observed particle has an uncertainty given approximately by

$$\Delta x \approx \frac{\lambda}{\sin \alpha}, \qquad (1.12)$$

where λ is the wavelength of the illumination and 2α is the angle subtended by the lens at the particle. We note that the resolution can be improved by reducing the wavelength of the radiation illuminating the particle; visible light waves are better than microwaves, and X-rays are better than visible light waves.

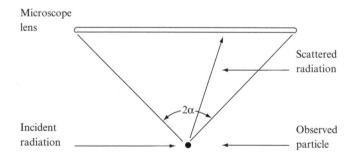

Fig. 1.6 A schematic illustration of the observation of a particle using the Heisenberg microscope. The particle scatters electromagnetic radiation with wavelength λ into a lens with angular aperture 2α and its position is determined with an uncertainty of $\Delta x \approx \lambda / \sin \alpha$.

However, because of the particle-like properties of light, the process of observation involves innumerable photon–particle collisions, with the scattered photons entering the lens of the microscope. To enter the lens, a scattered photon with wavelength λ and momentum h/λ must have a sideways momentum between

$$-\frac{h}{\lambda}\sin\alpha \quad \text{and} \quad +\frac{h}{\lambda}\sin\alpha.$$

Thus the sideways momentum of the scattered photon is uncertain to the degree

$$\Delta p \approx \frac{h}{\lambda}\sin\alpha. \tag{1.13}$$

The sideways momentum of the observed particle has a similar uncertainty, because momentum is conserved when the photon scatters.

We note that we can reduce the uncertainty in the momentum of the observed particle by increasing the wavelength of the radiation illuminating the particle, but this would result in a poorer spatial resolution of the microscope and an increase in the uncertainty in the position of the particle. Indeed, by combining Eq. (1.12) and Eq. (1.13), we find that the uncertainties in the position and in the momentum of the observed particle are approximately related by

$$\Delta x\,\Delta p \approx h. \tag{1.14}$$

This result is called the Heisenberg uncertainty principle. It asserts that greater accuracy in position is possible only at the expense of greater uncertainty in momentum, and vice versa. The precise statement of the principle is that the fundamental uncertainties in the simultaneous knowledge of the position and momentum of a particle obey the inequality

$$\Delta x\,\Delta p \geq \frac{\hbar}{2}, \quad \text{where } \hbar = \frac{h}{2\pi}. \tag{1.15}$$

We shall derive this inequality in Section 7.4 of Chapter 7.

The Heisenberg uncertainty principle suggests that a precise determination of position, one with $\Delta x = 0$, is possible at the expense of total uncertainty in momentum. In fact, an analysis of the microscope experiment, which takes into account the Compton effect, shows that a completely precise determination of position is impossible. According to the Compton effect, Eq. (1.2), the wavelength of a scattered photon is increased by

$$\Delta\lambda = \frac{h}{mc}(1 - \cos\theta),$$

where m is the mass of the observed particle and θ is an angle of scatter which will take the photon into the microscope lens. This implies that, even if we illuminate the particle with radiation of zero wavelength to get the best possible spatial resolution, the radiation entering the microscope lens has a wavelength of the order of h/mc. It follows that the resolution given by Eq. (1.12) is at best

$$\Delta x \approx \frac{\lambda}{\sin \alpha} \sim \frac{h}{mc \sin \alpha},\qquad (1.16)$$

which means that the minimum uncertainty in the position of an observed particle of mass m is of the order of h/mc.

Our analysis of Heisenberg's microscope experiment has illustrated the role of Planck's constant in a measurement: The minimum uncertainties in the position and momentum of an observed particle are related by $\Delta x \Delta p \approx h$, and the minimum uncertainty in position is not zero but of the order of h/mc. However, readers are warned that Heisenberg's microscope experiment can be misleading. In particular, readers should resist the temptation to believe that a particle can really have a definite position and momentum, which, because of the clumsy nature of the observation, cannot be measured. In fact, there is no evidence for the existence of particles with definite position and momentum. This concept is an unobservable idealization or a figment of the imagination of classical physicists. Indeed, the Heisenberg uncertainty principle can be considered as a danger signal which tells us how far we can go in using the classical concepts of position and momentum without getting into trouble with reality.

Measurement and wave–particle duality

In practice, the particle-like properties of a quantum particle are observed when it is detected, whereas its wave-like properties are inferred from the random nature of the observed particle-like properties. For example, in a two-slit experiment, particle-like properties are observed when the position of a quantum particle is measured on the screen, but the wave-like passage of the quantum particle through both the slits is not observed. It is inferred from a pattern of arrival at the screen which could only arise from the interference of two wave-like disturbances from the two slits.

However, the inferred properties of a quantum particle depend on the experiment and on the measurements that can take place in this experiment. We shall illustrate this subjective characteristic of a quantum particle by considering a modification of the two-slit experiment in which the screen can either be held fixed or be allowed to move as shown in Fig. 1.7.

When the pin in Fig. 1.7 is inserted, detectors on a fixed screen precisely measure the position of each arriving particle and an interference pattern builds with fringes separated by a distance of $\lambda D/d$.

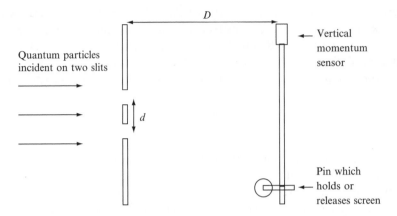

Fig. 1.7 A modified two-slit experiment in which the screen may move vertically and become a part of a detection system which identifies the slit through which each particle passes.

When the pin is withdrawn, the screen becomes a mobile detection system which is sensitive to the momentum $p = h/\lambda$ of the particles hitting the screen. It recoils when a particle arrives and, by measuring this recoil accurately, we can measure the vertical momentum of the particle detected at the screen and hence identify the slit from which the particle came. For example, near the centre of the screen, a particle from the upper slit has a downward momentum of $pd/2D$ and a particle from the lower slit has an upward momentum of $pd/2D$. In general, the difference in vertical momenta of particles from the two slits is approximately $\Delta p \approx pd/D$. Thus, if the momentum of the recoiling screen is measured with an accuracy of

$$\Delta p \approx \frac{pd}{D}, \tag{1.17}$$

we can identify the slit from which each particle emerges. When this is the case, a wave-like passage through both slits is not possible and an interference pattern should not build up. This statement can be verified by considering the uncertainties involved in the measurement of the momentum of the screen.

The screen is governed by the Heisenberg uncertainty principle and an accurate measurement of its momentum is only possible at the expense of an uncertainty in its position. In particular, if the uncertainty in the vertical momentum of the screen is $\Delta p \approx pd/D$, so that we can just identify the slit through which each particle passes, then the minimum uncertainty in the vertical position of the screen is

$$\Delta x \approx \frac{h}{\Delta p} \approx \frac{hD}{pd}. \tag{1.18}$$

This uncertainty in position can be rewritten in terms of the wavelength of the particle. Using $p = h/\lambda$, we obtain

$$\Delta x \approx \frac{\lambda D}{d}. \tag{1.19}$$

We note that this uncertainty in the vertical position of the screen is sufficient to wash out the interference fringes which would have a spacing of $\lambda D/d$. Hence, when the pin in Fig. 1.7 is withdrawn so that the recoiling screen can signal the slit from which a particle comes, no interference pattern builds up and no wave-like passage through both slits is inferred.

This thought experiment illustrates how the concepts of measurement and uncertainty can be used to provide a logical and consistent description of the wave-particle properties of quantum particles. In particular, it shows that, when it is possible to identify the slit through which a particle passes, there is no wave-like passage through both slits, but when there is no possibility of identifying the slit, the particle covertly passes through both slits in a wave-like way. In fact, the wave-like behaviour of quantum particles is always covert. Unlike a classical electromagnetic wave, the wave describing a quantum particle cannot be directly observed.[2]

Finally, some readers may find it instructively disturbing to consider a further variation of the two-slit experiment which was pointed out by Wheeler in 1978. In this variation, we imagine a situation in which the choice of the experimental arrangement in Fig. 1.7 is delayed until after the particle has passed the slits. We could, for example, insert the pin and fix the position of the screen just before each particle arrives at the screen. In this case an interference pattern builds up, which is characteristic of wave-like particles which pass through both slits. Alternatively, just before each particle arrives at the screen, we could withdraw the pin so as to allow the screen to recoil and determine the slit from which the particle comes. In this case, no interference pattern builds up.

Thus, a delayed choice of the experimental arrangement seems to influence the behaviour of the particle at an earlier time. The choice of arrangement seems to make history, either a history in which the particle passes through both slits or a history in which it passes through one or other of the slits. Before you dismiss this as unacceptable behaviour, note that the history created in this experiment is not classical history. The particles concerned are not classical

---

[2] A real experiment of this kind is described by Greenberger (*Reviews of Modern Physics* vol. 55, 1983). In this experiment polarized neutrons, that is neutrons with their spin orientated in a specific direction, pass through two slits. The polarization of the neutrons which pass through one of the slits is reversed, and then the intensity of neutrons with a specific polarization is measured at the screen. If the specific polarization at the screen is chosen so that one *cannot* infer through which slit each neutron passes, an interference pattern builds up. If it is chosen so that one *can* infer through which slit each neutron passes, no interference pattern builds up.

particles which pass through one slit or the other, nor are they classical waves which pass through both slits. They are quantum particles which have the capability to behave in both of these ways, but only one of these ways may be inferred in a particular experimental arrangement. The history created in a delayed choice experiment is an inferred history of a quantum particle.

Measurement and non-locality

The most important implication of this discussion of measurement is that quantum mechanics only describes what we can know about the world. For example, because we cannot know the position and momentum of an electron with precision, we cannot describe a world in which an electron has both a precise position and momentum. In the standard interpretation of quantum mechanics, a precise position or a precise momentum of an electron can be brought into existence by a measurement, but no attempt is made to explain how this occurs.

That properties are brought into existence by measurement is not restricted to the measurement of the position or momentum of a single particle. It applies to other observable properties of a quantum particle and also to systems of quantum particles. Amazingly systems of quantum particles exist in which a measurement at one location can bring into existence a property at a remote location. In other words a measurement *here* can affect things over *there*. Thus, measurements can have a non-local impact on our knowledge of the world.

The non-local nature of quantum mechanical measurement is best illustrated by considering a particular situation in which two photons are emitted by excited states of an atom. These photons may move off in opposite directions with the same polarization in the following meaning of the word *same*: If the photon moving to the East, say, is observed to have right-hand circular polarization, then the photon moving to the West is certain to be found to have right-hand circular polarization. But if the photon moving to the East is observed to have left-hand circular polarization, then the photon moving to the West is certain to be found to have left-hand circular polarization.

This behaviour would be unremarkable if right and left-hand polarization were two alternatives which were created at the moment the two photons were emitted. But this is not the case. At the moment of emission an *entangled state* is created in which the photons are simultaneously right and left-handed, but only one of these two alternatives is brought into existence by a subsequent measurement. Amazingly, when this measurement is performed on one of the photons, say the photon moving East, there are two outcomes: The *observed* photon moving to the East has a specific polarization and the *unobserved* photon moving West immediately has the same polarization.

This system of entangled photons is equivalent to having two ambidextrous gloves separated by a large distance but linked in such a way that if one glove becomes a right-hand glove, the other automatically becomes a right-hand glove. The quantum mechanical reasons for this unexpected togetherness of distant objects are that:

(1) the initial state of the gloves is a superposition of right and left-handedness, in much the same way as the state of a quantum particle can be a linear superposition of two waves passing through two slits; and

(2) a measurement not only disturbs what is measured but also brings into existence what is measured.

Needless to say, it is not possible to fully justify these arguments at the beginning of a book whose aim is to introduce the theory of quantum particles. But these arguments form a part of a logically consistent theory and they are supported by experimental observations, particularly by Alain Aspect and his colleagues; see, for example, *The Quantum Challenge* by G. Greenstein and A. G. Zajonc, (Jones and Bartlett, 1997).

The topics considered in this chapter provide the guidelines for a theory of quantum particles. Most importantly, the theory must provide a way of dealing with the particle and wave-like properties of quantum particles and in so doing it must involve the constant which links these properties, Planck's constant. In addition, the theory must recognize that measurement is not a passive act which has no effect on the observed system, but a way of creating a particular property of the system. The basic elements of such a theory will be developed in the next three chapters and then further developed by application in subsequent chapters. This development necessarily entails abstract mathematical concepts, but the results are not abstract because they describe what we can know about the world. However, we shall limit our dissussion to a world of non-relativistic particles Relativistic particles, like photons, will not be considered because this presents the additional challenge of dealing with the creation and destruction of particles.

PROBLEMS 1

1. Compton scattering can be described in terms of a collision between a photon and an electron, as shown in Fig. 1.1. The energy E and momentum P of a relativistic electron and the energy ϵ and momentum p of a photon are related by

$$E^2 - P^2 c^2 = m_e^2 c^4 \quad \text{and} \quad \epsilon = pc.$$

Let E_i, P_i and E_f, P_f denote the initial and final energies and momenta of the electron, and let ϵ_i, p_i and ϵ_f, p_f denote the initial and final energies and momenta of the photon. Assume that the electron is initially at rest, so that

$E_i = m_e c^2$ and $P_i = 0$, and assume that the photon is scattered through an angle θ.

By considering the conservation of momentum show that

$$\epsilon_i^2 - 2\epsilon_i\epsilon_f \cos\theta + \epsilon_f^2 = E_f^2 - m_e^2 c^4.$$

Write down the equation describing the conservation of relativistic energy in the collision and show that it may be rearranged to give

$$\epsilon_i^2 - 2\epsilon_i\epsilon_f + \epsilon_f^2 = E_f^2 - 2E_f m_e^2 c^4 + m_e^2 c^4.$$

By subtracting these equations show that

$$m_e c^2 \frac{(\epsilon_i - \epsilon_f)}{\epsilon_i\epsilon_f} = (1 - \cos\theta).$$

Show that this equation implies that the increase in wavelength of the scattered photon is given by

$$\Delta\lambda = \frac{h}{m_e c}(1 - \cos\theta).$$

2. In the photoelectric effect, electromagnetic radiation incident on a metal surface may eject electrons, but only if the frequency of the radiation exceeds a threshold value.

   Show that this frequency threshold can be understood if the mechanism for the photoelectric effect involves an interaction between a photon and an electron, and if the energy needed to eject the electron has a minimum value.

   The minimum energy needed to eject an electron from the surface of magnesium is 3.68 eV. Show that light with a frequency below 8.89×10^{14} Hz cannot produce photoelectrons from magnesium, no matter how intense the illumination may be.

3. Dimensional analysis can provide insight into Stefan–Boltzmann's law for the radiation from a black body. According to this law the intensity of radiation, in units of J s^{-1} m^{-2}, from a body at temperature T is

$$I = \sigma T^4,$$

where σ is Stefan–Boltzmann's constant. Because black-body radiation can be considered to be a gas of photons, i.e. quantum particles which move with velocity c with typical energies of the order of kT, the intensity I is a function of h, c and kT. Use dimensional analysis to confirm that I is proportional to T^4 and find the dependence of σ on h and c.

4. In Section 1.3 we used dimensional analysis to show that the size of a hydrogen atom can be understood by assuming that the electron in the atom is wave-like and non-relativistic. In this problem we show that, if we assume the electron in the atom is a classical electron described by the theory of relativity, dimensional analysis gives an atomic size which is four orders of magnitude too small.

   Consider a relativistic, classical theory of an electron moving in the Coulomb potential of a proton. Such a theory only involves three physical constants: m_e, $e^2/4\pi\epsilon_0$, and c, the maximum velocity in relativity. Show that it is possible to construct a length from these three physical constants, but show that it too small to characterize the size of the atom.

5. An electron in a circular orbit about a proton can be described by classical mechanics if its angular momentum L is very much greater than $\hbar$. Show that this condition is satisfied if the radius of the orbit r is very much greater than the Bohr radius a_0, i.e. if

$$r \gg a_0 = \frac{4\pi\epsilon_0}{e^2}\frac{\hbar^2}{m_e}.$$

6. Assume that an electron is located somewhere within a region of atomic size. Estimate the minimum uncertainty in its momentum. By assuming that this uncertainty is comparable with its average momentum, estimate the average kinetic energy of the electron.

7. Assume that a charmed quark of mass $1.5\,\text{GeV}/c^2$ is confined to a volume with linear dimension of the order of 1 fm. Assume that the average momentum of the quark is comparable with the minimum uncertainty in its momentum. Show that the confined quark may be treated as a non-relativistic particle, and estimate its average kinetic energy.

8. JJ and GP Thomson, father and son, both performed experiments with beams of electrons. In 1897, JJ deduced electrons are particles with a definite value for e/m_e. In 1927, GP deduced that electrons behave like waves. In JJ's experiment, electrons with kinetic energies of 200 eV passed through a pair of plates with 2 cm separation. Explain why JJ saw no evidence for wave-like behaviour of electrons.

9. The wave properties of electrons were first demonstrated in 1925 by Davisson and Germer at Bell Telephone Laboratories. The basic features of their experiment are shown schematically below.

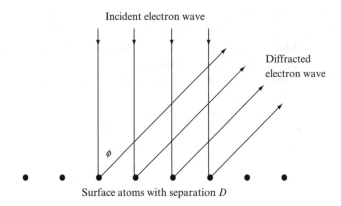

Electrons with energy 54 eV were scattered by atoms on the surface of a crystal of nickel.[3] The spacing between parallel rows of atoms on the surface was $D = 0.215\,\text{nm}$. Explain why Davisson and Germer detected strong scattering at an angle ϕ equal to 50 degrees.

10. The electrons which conduct electricity in copper have a kinetic energy of about 7 eV. Calculate their wavelength. By comparing this wavelength with the interatomic distance in copper, assess whether the wave-like properties of conduction electrons are important as they move in copper.

    (The density of copper is $8.9 \times 10^3\,\text{kg m}^{-3}$ and the mass of a copper atom is 60 amu.)

11. Neutrons from a nuclear reactor are brought into thermal equilibrium by repeated collisions in heavy water at $T = 300\,\text{K}$. What is the average energy (in eV) and the typical wavelength of the neutrons? Explain why they are diffracted when they pass through a crystalline solid.

12. Estimate the wavelength of an oxygen molecule in air at NTP. Compare this wavelength with the average separation between molecules in air and explain why the motion of oxygen molecules in air at NTP is not affected by the wave-like properties of the molecules.

---

[3] Incidentally, the crystalline structure was caused by accident when they heated the target in hydrogen in an attempt to repair the damage caused by oxidation.

2

The Schrödinger equation

The first step in the development of a logically consistent theory of non-relativistic quantum mechanics is to devise a wave equation which can describe the covert, wave-like behaviour of a quantum particle. This equation is called the Schrödinger equation.

The role of the Schrödinger equation in quantum mechanics is analogous to that of Newton's Laws in classical mechanics. Both describe motion. Newton's Second Law is a differential equation which describes how a classical particle moves, whereas the Schrödinger equation is a partial differential equation which describes how the wave function representing a quantum particle ebbs and flows. In addition, both were postulated and then tested by experiment.

2.1 WAVES

As a prelude to the Schrödinger equation, we shall review how mathematics can be used to describe waves of various shapes and sizes.

Sinusoidal waves

The most elegant wave is a sinusoidal travelling wave with definite wavelength λ and period τ, or equivalently definite wave number, $k = 2\pi/\lambda$, and angular frequency, $\omega = 2\pi/\tau$. Such a wave may be represented by the mathematical function

$$\Psi(x, t) = A \cos(kx - \omega t) \tag{2.1}$$

where A is a constant. At each point x, the function $\Psi(x, t)$ oscillates with amplitude A and period $2\pi/\omega$. At each time t, the function $\Psi(x, t)$ undulates with amplitude A and wavelength $2\pi/k$. Moreover, these undulations move, like a Mexican wave, in the direction of increasing x with velocity ω/k; for example, the maximum of $\Psi(x, t)$ corresponding to $kx - \omega t = 0$ occurs at the position $x = \omega t/k$, and the minimum corresponding to $kx - \omega t = \pi$ occurs at the position $x = \lambda/2 + \omega t/k$; in both cases the position moves with velocity ω/k.

The function $\sin(kx - \omega t)$, like $\cos(kx - \omega t)$, also represents a sinusoidal travelling wave with wave number k and angular frequency ω. Because

$$\sin(kx - \omega t) = \cos(kx - \omega t - \pi/2),$$

the undulations and oscillations of $\sin(kx - \omega t)$ are out of step with those of $\cos(kx - \omega t)$; the waves $\sin(kx - \omega t)$ and $\cos(kx - \omega t)$ are said to have a phase difference of $\pi/2$. The most general sinusoidal travelling wave with wave number k and angular frequency ω is the linear superposition

$$\Psi(x, t) = A\cos(kx - \omega t) + B\sin(kx - \omega t), \tag{2.2}$$

where A and B are arbitrary constants.

Very often in classical physics, and invariably in quantum physics, sinusoidal travelling waves are represented by complex exponential functions of the form

$$\Psi(x, t) = A\,e^{i(kx - \omega t)}. \tag{2.3}$$

The representation of waves by complex exponentials in classical physics is merely a mathematical convenience. For example, the pressure in a sound wave may be described by the real function $A\cos(kx - \omega t)$, but this real function may be taken to be the real part of a complex exponential function $A\,e^{i(kx - \omega t)}$ because

$$e^{i(kx - \omega t)} = \cos(kx - \omega t) + i\sin(kx - \omega t).$$

Thus, in classical physics, we have the option of representing a real sinusoidal wave by the real part of a complex exponential. In quantum physics, however, the use of complex numbers is not an option and we shall see that a complex exponential provides a natural description of a de Broglie wave.

Linear superpositions of sinusoidal waves

Two sinusoidal waves moving in opposite directions may be combined to form standing waves. For example, the linear superposition

$$A\cos(kx - \omega t) + A\cos(kx + \omega t)$$

gives rise to the wave $2A\cos kx \cos \omega t$. This wave oscillates with period $2\pi/\omega$ and undulates with wavelength $2\pi/k$, but these oscillations and undulations do not propagate; it is a non-Mexican wave which merely stands and waves.

Alternatively, many sinusoidal waves may be combined to form a wave packet. For example, the mathematical form of a wave packet formed by a linear superposition of sinusoidal waves with constant amplitude A and wave numbers in the range $k - \Delta k$ to $k + \Delta k$ is

$$\Psi(x, t) = \int_{k-\Delta k}^{k+\Delta k} A\cos(k'x - \omega't)\, \mathrm{d}k'. \tag{2.4}$$

If k is positive, this wave packet travels in the positive x direction, and in the negative x direction if k is negative.

The initial shape of the wave packet, i.e. the shape at $t = 0$, may be obtained by evaluating the integral

$$\Psi(x, 0) = \int_{k-\Delta k}^{k+\Delta k} A\cos k'x\, \mathrm{d}k'.$$

This gives

$$\Psi(x, 0) = S(x)\cos kx, \qquad \text{where} \quad S(x) = 2A\Delta k \frac{\sin(\Delta k x)}{(\Delta k x)}. \tag{2.5}$$

If $\Delta k \ll k$, we have a rapidly varying sinusoidal, $\cos kx$, with an amplitude modulated by a slowly varying function $S(x)$ which has a maximum at $x = 0$ and zeros when x is an integer multiple of $\pi/\Delta k$. The net result is a wave packet with an effective length of about $2\pi/\Delta k$. Three such wave packets, with different values for Δk, are illustrated in Fig. 2.1. We note that the wave packets increase in length as the range of wave numbers decreases and that they would become 'monochromatic' waves of infinite extent as $\Delta k \to 0$. Similar behaviour is exhibited by other types of wave packets.

The velocity of propagation of a wave packet, and the possible change of shape as it propagates, depend crucially on the relation between the angular frequency and wave number. This relation, the function $\omega(k)$, is called the dispersion relation because it determines whether the waves are dispersive or non-dispersive.

Dispersive and non-dispersive waves

The most familiar example of a non-dispersive wave is an electromagnetic wave in the vacuum. A non-dispersive wave has a dispersion relation of the form $\omega = ck$, where c is a constant so that the velocity of a sinusoidal wave, $\omega/k = c$,

(A)

(B)

(C)

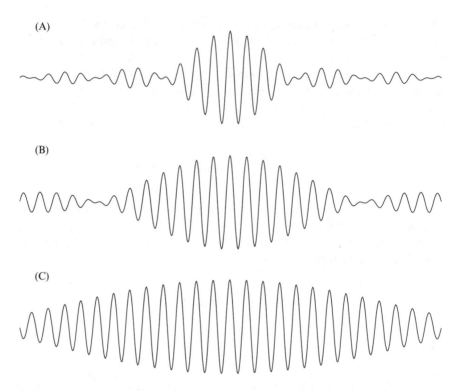

Fig. 2.1 The initial shapes of the wave packets given by a linear superposition of sinusoidal waves with constant amplitude A and wave numbers in the range $k - \Delta k$ to $k + \Delta k$; see Eq. (2.5). The three diagrams show how the length of a wave packet increases as the range of wave numbers Δk decreases. The value of $A\Delta k$ is constant, but Δk equals $k/8$ in diagram (A), Δk equals $k/16$ in diagram (B) and Δk equals $k/32$ in diagram (C). In general, the length of a wave packet is inversely proportional to Δk and becomes infinite in extent as $\Delta k \to 0$.

is independent of the wave number k. A wave packet formed from a linear superposition of such sinusoidal waves travels without change of shape because each sinusoidal component has the same velocity.

Non-dispersive waves are governed by a partial differential equation called the classical wave equation. For waves travelling in three dimensions, it has the form

$$\nabla^2 \Psi - \frac{1}{c^2}\frac{\partial^2 \Psi}{\partial t^2} = 0, \quad \text{where} \quad \nabla^2 = \frac{\partial^2}{\partial x^2} + \frac{\partial^2}{\partial y^2} + \frac{\partial^2}{\partial z^2}, \tag{2.6}$$

and for waves travelling in one dimension, the x direction say, it has the form

$$\frac{\partial^2 \Psi}{\partial x^2} - \frac{1}{c^2}\frac{\partial^2 \Psi}{\partial t^2} = 0. \tag{2.7}$$

The classical wave equation has an infinite number of solutions corresponding to an infinite variety of wave forms. For example, the sinusoidal waves,

$$A\cos{(kx - \omega t)}, \qquad A\sin{(kx - \omega t)}, \qquad \text{or } A\,e^{i(kx-\omega t)},$$

are solutions provided $\omega^2 = c^2 k^2$, as may be shown by direct substitution into Eq. (2.7); solutions with $k = +\omega/c$ describe waves travelling in the positive x direction and solutions with $k = -\omega/c$ describe waves travelling in the negative x direction. Because each term in the classical wave equation is linear in Ψ, a linear superposition of sinusoidal waves is also a solution. For example, a superposition like Eq. (2.4) is a solution which describes a wave packet which propagates without change of shape.

However, the majority of waves encountered in classical and in quantum physics are dispersive waves. A dispersive wave is governed by a partial differential equation which is more complicated than the classical wave equation, Eq. (2.7).[1] The dispersion relation is more complicated than $\omega = ck$ so that the velocity of propagation of a sinusoidal wave, ω/k, depends upon the wave number k. Hence a packet of dispersive waves will, in general, change shape as it propagates. However, if the packet is composed of waves with a narrow range of wave numbers, it has a well-defined velocity of propagation. This velocity is called the *group velocity* and it is given by

$$v_{group} = \frac{d\omega}{dk}, \tag{2.8}$$

whereas the velocity of a simple sinusoidal wave, ω/k, is called the *phase velocity*.

To understand Eq. (2.8), we note that the group velocity describes the motion of a localized disturbance due to constructive interference of many sinusoidal waves. Let us focus on the point of constructive interference of two sinusoidal waves with wave numbers k_1 and k_2 and angular frequencies ω_1 and ω_2 which is formed when the waves are in phase; i.e. when

$$k_1 x - \omega_1 t = k_2 x - \omega_2 t.$$

By rearranging this equation, we find that the position of this point of constructive interference is given by

$$x = \left(\frac{\omega_1 - \omega_2}{k_1 - k_2}\right) t.$$

---

[1] In fact, Eq. (2.7) with the constant c replaced by a frequency-dependent velocity is often used to describe dispersive waves.

Thus our point of constructive interference is located at $x = 0$ when $t = 0$ and it moves with a velocity given by $(\omega_1 - \omega_2)/(k_1 - k_2)$, or by Eq. (2.8) if $|k_1 - k_2|$ is small. Of course, with two sinusoidal waves, there are an infinite number of points of constructive interference, but many sinusoidal waves can form a localized region of constructive interference which moves with a velocity given by Eq. (2.8).

To illustrate how a group velocity can be derived from Eq. (2.8), we consider the example of water waves of long wavelength which obey the dispersion relation

$$\omega = \sqrt{gk},$$

where g is the acceleration due to gravity. The velocity of a sinusoidal water wave, the so-called phase velocity, is

$$v_{phase} = \frac{\omega}{k} = \sqrt{\frac{g}{k}},$$

and the velocity of a packet of water waves with a narrow range of wave numbers near k is

$$v_{group} = \frac{d\omega}{dk} = \frac{1}{2}\sqrt{\frac{g}{k}}.$$

Thus, for water waves, the group velocity is exactly one-half of the phase velocity. In other words, the sinusoidal waves forming the packet, travel at twice the speed of the region of maximum disturbance formed by the interference of these waves. However, the shape of the disturbance will change as it propagates; in general it will tend to spread out.

2.2   PARTICLE WAVE EQUATIONS

In classical physics, fundamental laws of physics are used to derive the wave equations which describe wave-like phenomena; for example, Maxwell's laws of electromagnetism can be used to derive the classical wave equation (2.6) which governs electromagnetic waves in the vacuum. In contrast, we shall view the wave equation governing the wave-like properties of a quantum particle as a fundamental equation which cannot be derived from underlying basic physical principles. We, like the inventors of quantum theory, can only guess the form of this wave equation and then test for consistency and agreement with experiment.

A wave equation for a free particle

We shall construct a possible wave equation for a freely moving non-relativistic particle by considering the properties of the de Broglie waves describing the particle.

According to Eq. (1.3) a particle with momentum p has a de Broglie wavelength given by $\lambda = h/p$. This implies that a de Broglie wave with wave number $k = 2\pi/\lambda$ describes a particle with momentum

$$p = \hbar k, \quad \text{where} \quad \hbar = \frac{h}{2\pi}. \tag{2.9}$$

We shall extend this idea by assuming that a de Broglie wave packet with a range of wave numbers between $k - \Delta k$ and $k + \Delta k$ describes a particle with an uncertain momentum

$$\Delta p \approx \hbar \Delta k. \tag{2.10}$$

We shall also assume that the length of this wave packet is a measure of Δx, the uncertainty in the position of the particle. Using Eq. (2.5) and Fig. 2.1 as a guide, we write

$$\Delta x \approx \frac{2\pi}{\Delta k}. \tag{2.11}$$

If we multiply these uncertainties, we obtain

$$\Delta x\, \Delta p \approx h,$$

in agreement with the Heisenberg uncertainty principle, Eq. (1.14). Thus, a de Broglie wave packet can account for the uncertainties in the position and momentum of a quantum particle.

However, we note that a de Broglie wave must be transformed by a measurement. If a precise measurement of the position is made, the new wave packet describing the particle must be very short, a superposition of sinusoidal waves with a very wide range of wavelengths. Similarly, if a precise measurement of the momentum is made, the new wave packet is very long with a sharply defined wavelength. This implies that the wave packet is a fragile entity which is transformed by a measurement. No one knows how this happens.

We shall now impose the condition that the wave packet represents a moving quantum particle. Specifically, we shall require that the group velocity of the packet is equal to the velocity of a particle with mass m and momentum $p = \hbar k$; i.e. we shall require that

$$\frac{d\omega}{dk} = \frac{\hbar k}{m}.$$ (2.12)

This equation may be integrated to give the following dispersion relation for the de Broglie waves describing a freely moving quantum particle of mass m:

$$\omega = \frac{\hbar k^2}{2m}.$$ (2.13)

In obtaining this relation we have set the constant of integration to zero because this constant gives rise to no observable consequences in non-relativistic quantum mechanics.

Our task is to find a wave equation which has sinusoidal solutions which obey this dispersion relation. The simplest such wave equation is called the Schrödinger equation. For a free particle moving in one dimension, it has the form

$$i\hbar \frac{\partial \Psi}{\partial t} = -\frac{\hbar^2}{2m} \frac{\partial^2 \Psi}{\partial x^2}.$$ (2.14)

It is easy to verify that the complex exponential

$$\Psi(x, t) = A\, e^{i(kx - \omega t)}$$ (2.15)

is a solution of this equation provided ω and k obey the dispersion relation Eq. (2.13). If we substitute into Eq. (2.14), the left-hand side yields

$$i\hbar \frac{\partial \Psi}{\partial t} = i\hbar(-i\omega)A\, e^{i(kx - \omega t)} = \hbar \omega A\, e^{i(kx - \omega t)},$$

and the right-hand side yields

$$-\frac{\hbar^2}{2m} \frac{\partial^2 \Psi}{\partial x^2} = \frac{\hbar^2 k^2}{2m} A\, e^{i(kx - \omega t)},$$

and we have a solution provided $\hbar \omega = \hbar^2 k^2 / 2m$.

Because the sinusoidal solution, Eq. (2.15), describes a wave moving in the x direction with wave number k and angular velocity ω, we shall assume that it represents a free particle moving in the x direction with a sharply defined momentum $p = \hbar k$ and energy $E = p^2/2m = \hbar \omega$. There are, of course, many other solutions of the Schrödinger equation which represent other states of motion of the particle.

We emphasize that in order to accommodate the dispersion relation for de Broglie waves, Eq. (2.13), we have arrived at a wave equation, the free-particle

Schrödinger equation Eq. (2.14), whose solutions are necessarily complex functions of space and time. These complex functions are called wave functions. We recall that classical waves are often represented by complex functions, but this representation is purely a matter of mathematical convenience; classical waves are real functions of space and time. In contrast, Schrödinger wave functions are not real functions of space and time. They are complex functions which describe the covert wave-like behaviour of a quantum particle.

So far we have only considered sinusoidal solutions of the Schrödinger equation, but given these solutions we can construct other types of solutions. Because each term in the Schrödinger equation is linear in the wave function Ψ, a superposition of solutions is also a solution. For example,

$$\Psi(x, t) = A_1 \, e^{i(k_1 x - \omega_1 t)} + A_2 \, e^{i(k_2 x - \omega_2 t)},$$

with

$$\hbar\omega_1 = \frac{\hbar^2 k_1^2}{2m} \quad \text{and} \quad \hbar\omega_2 = \frac{\hbar^2 k_2^2}{2m},$$

where A_1 and A_2 are arbitrary complex constants, is a solution of Eq. (2.14); this may easily be confirmed by direct substitution. Indeed, the most general solution is a superposition of sinusoidal waves with all possible angular frequencies and wave numbers; i.e.

$$\Psi(x, t) = \int_{-\infty}^{+\infty} A(k') \, e^{i(k' x - \omega' t)} \, dk' \quad \text{with} \quad \hbar\omega' = \frac{\hbar^2 k'^2}{2m}. \tag{2.16}$$

In this superposition, $A(k')$ is an arbitrary complex function of k' and the integral represents a sum over all possible values of k'. If the function $A(k')$ is such that the sum involves a narrow range of wave numbers around a positive value k, this superposition yields a wave packet moving in the positive x direction with a group velocity $\hbar k/m$. Such a wave packet represents a quantum particle which moves with velocity $\hbar k/m$, but with position and momentum in accordance with the Heisenberg uncertainty principle.

Wave equation for a particle in a potential energy field

The interactions of a non-relativistic particle can usually be described in terms of a potential energy field. For example, an electron in a hydrogen atom can be thought of as moving in the potential energy field $V(r) = -e^2/4\pi\epsilon_0 r$ of a nucleus. In classical mechanics, this field implies that an electron at a distance r from the nucleus experiences an attractive force of magnitude $e^2/4\pi\epsilon_0 r^2$. In quantum mechanics, it implies that the wave equation for the electron is not the simple free-particle wave equation given by Eq. (2.14).

In 1926, Erwin Schrödinger invented a wave equation for a quantum particle in a potential energy field which led to a successful description of atoms and other microscopic systems. It is a generalization of the wave equation for a free particle given by Eq. (2.14). The Schrödinger equation for a particle moving in the three-dimensional potential energy field $V(\mathbf{r})$ is

$$i\hbar\frac{\partial \Psi}{\partial t} = \left[-\frac{\hbar^2}{2m}\nabla^2 + V(\mathbf{r})\right]\Psi. \tag{2.17}$$

When the particle moves in a one-dimensional potential $V(x)$ the Schrödinger equation simplifies to

$$i\hbar\frac{\partial \Psi}{\partial t} = \left[-\frac{\hbar^2}{2m}\frac{\partial^2}{\partial x^2} + V(x)\right]\Psi. \tag{2.18}$$

It is easy to find solutions of the Schrödinger equation when the potential energy is a constant. For example, when a particle moves along the x axis with constant potential energy V_0, the wave function

$$\Psi(x, t) = A\, e^{i(kx-\omega t)}$$

is a solution of Eq. (2.18) provided

$$\hbar\omega = \frac{\hbar^2 k^2}{2m} + V_0.$$

This wave function represents a particle with sharply defined total energy E and momentum p given by

$$E = \frac{p^2}{2m} + V_0 \quad \text{and} \quad p = \hbar k.$$

In later chapters we shall find solutions of the Schrödinger equation for a variety of potential energy fields, including the most important case of the Coulomb potential energy of an electron in an atom. But our next task is to explore the meaning of the Schrödinger equation.

PROBLEMS 2

1. Waves on the surface of water are dispersive waves. If the wavelength is short so that surface tension provides the restoring force, the dispersion relation is

$$\omega = \sqrt{\frac{Tk^3}{\rho}},$$

where T is the surface tension and ρ is the density of water.

   Find expressions for the phase velocity of a sinusoidal water wave with wave number k and the group velocity of a packet of water waves with wave numbers near k.

2. Consider the wave packet represented by

$$\Psi(x, t) = \int_{k-\Delta k}^{k+\Delta k} A \cos (k'x - \omega' t) \, dk'$$

where A is a constant and assume that the dispersion relation is

$$\omega' = ck',$$

where c is a constant. By integrating over k' show that

$$\Psi(x, t) = S(x - ct) \cos k(x - ct)$$

where

$$S(x - ct) = 2A\Delta k \frac{\sin [\Delta k(x - ct)]}{[\Delta k(x - ct)]}.$$

Describe the propagation properties of this wave packet.

3. Verify by direct substitution that the real functions

$$\Psi = A \cos (kx - \omega t) \quad \text{and} \quad \Psi = A \sin (kx - \omega t)$$

are not solutions of the Schrödinger equation for a free particle.

4. Verify that the wave function

$$\Psi(x, t) = A e^{i(kx - \omega t)} - A e^{-i(kx + \omega t)},$$

where A is an arbitrary complex constant, is a solution of the Schrödinger equation for a free particle of mass m, Eq. (2.14), if

$$\hbar\omega = \frac{\hbar^2 k^2}{2m}.$$

Show that this wave function can be rewritten as

$$\Psi(x, t) = 2iA \sin kx\, e^{-i\omega t}.$$

What sort of wave is this?

5. In quantum mechanics it is the convention to represent a free particle with momentum p and energy E by the wave function

$$\Psi(x, t) = A\, e^{+i(px-Et)/\hbar}.$$

Physicists on another planet may have chosen the convention of representing a free particle with momentum p and energy E by the wave function

$$\Psi(x, t) = A\, e^{-i(px-Et)/\hbar}.$$

What is the form of the Schrödinger equation on this planet?

6. This question tentatively considers the wave equation for a relativistic particle. According to the Theory of Relativity, the momentum p and energy ϵ of a particle of mass m are related by

$$\epsilon^2 - p^2 c^2 = m^2 c^4,$$

and the velocity of the particle is given by

$$v = \frac{pc^2}{\epsilon}.$$

(a) Assume that the motion of the particle can be described by a wave packet in which the angular frequency and wave number are given by

$$\epsilon = \hbar\omega \quad \text{and} \quad p = \hbar k.$$

Derive an expression for the group velocity of the wave packet and show that it is identical to the particle velocity.

(b) Show that the wave equation

$$\frac{\partial^2 \Psi}{\partial t^2} - c^2 \frac{\partial^2 \Psi}{\partial x^2} + \frac{m^2 c^4}{\hbar^2} \Psi = 0$$

has solutions of the form

$$\Psi = A\,e^{-i(\omega t - kx)},$$

which could possibly describe a relativistic particle of mass m with energy $\epsilon = \hbar\omega$ and momentum $p = \hbar k$.

(At first sight this wave equation, which is called the Klein–Gordon equation, provides a basis for a theory for relativistic particles which is similar to that provided by the Schrödinger equation for non-relativistic particles. However, a sensible interpretation of the Klein–Gordon equation cannot be obtained if its solutions are treated as wave functions. A sensible interpretation only emerges if the solutions are treated as quantum field operators.)

3

Position and momentum

The Schrödinger equation has an infinite number of solutions which correspond to an infinite number of possible states of motion. These wave functions, being extended and non-localized, can easily describe the wave-like properties of a particle, but it is difficult to see how they can describe particle-like properties. In a two-slit experiment, for example, a wave function can describe how an electron can pass through two slits, but how can it describe an electron that arrives as a lump on the screen? This problem may be resolved by adopting the radical idea that measurements can lead to random outcomes governed by the laws of probability.

In this chapter we shall focus on how the uncertain outcomes of position and momentum measurements are described in quantum mechanics. We shall show that these outcomes are governed by the wave function of a particle. In so doing, we shall illustrate how a wave function provides a description of the properties that could emerge from measurements, in other words, a description of potentialities that may become realities.

3.1 PROBABILITY

Because of the importance of probability in quantum measurement, we shall briefly consider how discrete and continuous random variables are governed by probability distributions. These general considerations are illustrated by problems at the end of the chapter on the Poisson, Gaussian and exponential probability distributions.

Discrete random variables

Let us consider a process or experiment with possible outcomes described by a discrete random variable which can take on the values $x_0, x_1, x_2, \ldots$, with probabilities $p_0, p_1, p_2, \ldots$. The set of probabilities p_n is called a probability distribution. Because the total probability of all the possible outcomes is equal to one, the probability distribution p_n must satisfy the normalization condition

$$\sum_{\text{all } n} p_n = 1. \tag{3.1}$$

For example, if p_n is the probability that a reader of this book has n grandparents that are alive and well, then

$$p_0 + p_1 + p_2 + p_3 + p_4 = 1.$$

The probability distribution p_n can be used to evaluate the expectation value for the random variable x_n. This is the average value of the many possible outcomes that may occur when the process or experiment takes place an infinite number of times. It is given by

$$\langle x \rangle = \sum_{\text{all } n} x_n p_n. \tag{3.2}$$

The likely spread in the outcomes about this expectation value is given by the standard deviation or uncertainty in x. We shall denote this by Δx. The square of the standard deviation is called the variance and the variance is given by

$$(\Delta x)^2 = \sum_{\text{all } n} (x_n - \langle x \rangle)^2 p_n. \tag{3.3}$$

In this expression $(x_n - \langle x \rangle)$ is the deviation of x_n from the expected value; this deviation may be positive or negative and its average value is zero. However, the variance is the average of the square of this deviation; it is zero when there is only one possible outcome and it is a positive number when there is more than one possible outcome.

In many cases it is useful to rewrite Eq. (3.3) in the following way. Using

$$(x_n - \langle x \rangle)^2 = x_n^2 - 2\langle x \rangle x_n + \langle x \rangle^2,$$

and bearing in mind that $\langle x \rangle$ is a number that does not depend on n, we find

$$(\Delta x)^2 = \sum_{\text{all } n} x_n^2 p_n - 2\langle x \rangle \sum_{\text{all } n} x_n p_n + \langle x \rangle^2 \sum_{\text{all } n} p_n.$$

The first term is equal to $\langle x^2 \rangle$, the expectation value of the square of x_n. The second term is equal to $-2\langle x \rangle \langle x \rangle = -2\langle x \rangle^2$. The third term is equal to $\langle x \rangle^2$

because the probability distribution is normalized. It follows that the variance given by Eq. (3.3) may be rewritten as

$$(\Delta x)^2 = \langle x^2 \rangle - \langle x \rangle^2. \tag{3.4}$$

This equation states that the variance is the difference between the average of the square and the square of the average of the random variable.

Continuous random variables

We shall now consider a process or experiment in which the outcomes are described by a continuous variable x. The probability of an outcome between x and $x + dx$ can be denoted by $\rho(x)\, dx$. The function $\rho(x)$ is called a probability density. It satisfies the normalization condition

$$\int_{all\ x} \rho(x)\, dx = 1, \tag{3.5}$$

because the probability of an outcome x anywhere in its range of possible values must be equal to one. For example, if x is the position of a particle confined to the region $0 \le x \le a$, then

$$\int_0^a \rho(x)\, dx = 1.$$

The expectation value of x, in analogy with Eq. (3.2), is given by the integral

$$\langle x \rangle = \int_{all\ x} x\rho(x)\, dx. \tag{3.6}$$

Similarly, the expectation value of x^2 is given by

$$\langle x^2 \rangle = \int_{all\ x} x^2\rho(x)\, dx \tag{3.7}$$

and the standard deviation, or uncertainty, in x is given by

$$\Delta x = \sqrt{\langle x^2 \rangle - \langle x \rangle^2}. \tag{3.8}$$

3.2 POSITION PROBABILITIES

We shall now return to quantum mechanics and consider how the position of a quantum particle is described. This problem was first successfully addressed by Max Born by introducing an interpretation of the Schrödinger wave function which led to a revolution in the philosophical basis of physics. We shall introduce Born's interpretation by reconsidering the two-slit interference experiment discussed in Chapter 1. We shall describe how an interference pattern arises when a classical wave passes through two slits and then develop a way of describing how a similar pattern is produced by a current of quantum particles.

Two-slit interference

The key elements of a two-slit interference experiment are very simple: a source of something with wave-like properties, two slits and an observation screen, as illustrated in Fig. 3.1.

When a classical wave passes through the slits, two waves emerge which combine and interfere to form a pattern on the screen. These waves can be described by real functions of space and time. For example, if the wave number is k and the angular frequency is ω, the combined wave at a point P, at distance R_1 from slit S_1 and R_2 from slit S_2, may be represented by

$$\Psi = A_1 \cos(kR_1 - \omega t) + A_2 \cos(kR_2 - \omega t), \tag{3.9}$$

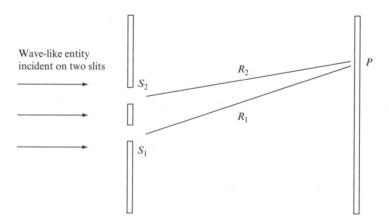

Fig. 3.1 A two-slit interference experiment in which a wave-like entity passes through two slits, S_1 and S_2, and is detected on a screen. Equally spaced bright and dark fringes are observed when wave-like disturbances from the two slits interfere constructively and destructively on the screen. Constructive interference occurs at the point P when the path difference $R_1 - R_2$ is an integer number of wavelengths.

where A_1 and A_2 are amplitudes which are inversely proportional to R_1 and R_2, respectively. The energy density and intensity of the wave at point P are proportional to the square of the wave. If we set $A_1 = A_2 = A$, which is a good approximation when the distances R_1 and R_2 are large compared with the slit separation, and if we use some simple trigonometry, we find[1]

$$\Psi^2 = 2A^2 \cos^2\left(\frac{k(R_1 - R_2)}{2}\right) \cos^2 \omega t. \tag{3.10}$$

It follows that the intensity has maxima when $k(R_1 - R_2)/2$ is an integer multiple of π and minima when it is a half-integer multiple of π. If we rewrite k in terms of the wavelength λ, we find that the maxima arise when the path difference $R_1 - R_2$ is equal to an integer number of wavelengths and the minima when $R_1 - R_2$ is equal to a half-integer number of wavelengths. When the effect of the finite width of the slits is taken into account, these maxima and minima give rise to an interference pattern like that illustrated in Fig. 3.2.

We shall now seek a similar description of how an interference pattern arises when a current of quantum particles passes through two slits. We shall assume that the covert passage of a quantum particle through the two slits is

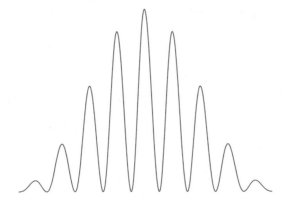

Fig. 3.2 An interference pattern produced by a classical wave and by a current of quantum particles passing through two slits with a finite width. Constructive interference occurs at points when the difference in paths from the slits equals an integer number of wavelengths; the central maximum occurs when the path difference is zero. Note the pattern given by a classical wave oscillates with time with a definite frequency whereas the pattern given by a current of quantum particles builds up gradually as more particles arrive at the screen.

---

[1] This trigonometry can be avoided by representing classical waves as the real part of a complex exponential, but we have deliberately not done this because we wish to emphasise that classical waves, unlike quantum wave functions, are real functions of space and time.

represented by a wave function, i.e. by a complex function of space and time that is a solution of the Schrödinger equation. If the particle has definite momentum $p = \hbar k$ and energy $E = \hbar \omega$, the wave function at a point P is a linear superposition of two terms, a wave from slit S_1 and a wave from slit S_2, of the form

$$\Psi = A_1 e^{+i(kR_1 - \omega t)} + A_2 e^{+i(kR_2 - \omega t)}, \tag{3.11}$$

where A_1 and A_2 are complex constants with approximately the same value.

When the particle arrives at the screen a very complicated process occurs. At each point on the screen there is a measuring device which magnifies microscopic effects due to the particle so that there is a clear signal that a particle has or has not arrived at that point. In other words, a small-scale event triggers something that can actually be seen. In practice, the particle may be detected at any point on the screen and, as more and more particles pass through the slits, an interference pattern builds up on the screen, as illustrated in Fig. 1.3.

We shall not attempt to understand how this complicated process of detection takes place. Instead, we shall set a more modest objective of describing the possible outcomes of the process using the Schrödinger wave function, Ψ. This can be achieved by making the bold assumption that the probability of detecting a particle at a particular location is proportional to the value of the *effective intensity* of the complex wave function at that location. In analogy with classical waves, we shall define this intensity to be a real number given by

$$|\Psi|^2 = \Psi^*\Psi. \tag{3.12}$$

The value of $|\Psi|^2$ at a point P on the screen can be found using the wave function Eq. (3.11). Making the approximation $A_1 = A_2 = A$, we have

$$|\Psi|^2 = (A^* e^{-i(kR_1 - \omega t)} + A^* e^{-i(kR_2 - \omega t)})(A e^{+i(kR_1 - \omega t)} + A e^{+i(kR_2 - \omega t)}),$$

which gives

$$|\Psi|^2 = |A|^2 + |A|^2 + |A|^2 e^{+ik(R_1 - R_2)} + |A|^2 e^{-ik(R_1 - R_2)}.$$

Using the mathematical relations

$$\cos\theta = (e^{+i\theta} + e^{-i\theta})/2 \quad \text{and} \quad \cos\theta = 2\cos^2\left(\frac{\theta}{2}\right) - 1,$$

we find

$$|\Psi|^2 = 2A^2 \cos^2\left(\frac{k(R_1 - R_2)}{2}\right). \tag{3.13}$$

If we compare with Eq. (3.10), we see that $|\Psi|^2$, has maximum and minimum values on the screen similar to those given by the intensity of a classical wave; there are maxima when $R_1 - R_2$ is an integer number of de Broglie wavelengths and minima when $R_1 - R_2$ is a half-integer number of de Broglie wavelengths. Hence, if the probability of detection is proportional to $|\Psi|^2$, an interference pattern similar to that shown in Fig. 3.2 will build up when many quantum particles pass through the two slits.

Thus, we have a logical way of describing the interference pattern produced by quantum particles. It is based on two crucial ideas. First, the wave function of the particle at the screen, Ψ, is a linear superposition of wave functions: a wave from slit S_1 and a wave from slit S_2. Second, the probability of detecting the particle at a particular location is proportional to the value of $|\Psi|^2$ at that location.

But we also need a successful way of describing the outcome of the modified two-slit experiment shown in Fig. 1.7. In particular, we need to explain why there is no interference pattern when we can identify the slit through which each particle passes. To do so, we shall assume that the process of identification changes the wave function of the particle, so that it becomes a single wave from the slit through which the particle passes. In so doing, we follow the standard practice of assuming that a measurement can affect a wave function. It is also standard practice not to explore too deeply how this actually occurs.

The Born interpretation of the wave function

The idea that the wave function can govern the potential outcomes of a position measurement was first proposed by Max Born in 1926. It is now called the Born interpretation of the wave function. According to this interpretation, the wave function $\Psi(\mathbf{r}, t)$ is a complex function of the space coordinates whose modulus squared, $|\Psi(\mathbf{r}, t)|^2$, is a measure of the probability of finding the particle at the point $\mathbf{r}$ at time t. The particle can be found anywhere, but it is more likely to be found at a point where $|\Psi(\mathbf{r}, t)|^2$ is large. Specifically

$$|\Psi(\mathbf{r}, t)|^2 \, \mathrm{d}^3\mathbf{r} = \left\{ \begin{array}{l} \text{The probability of finding the particle} \\ \text{at time t in the volume element } \mathrm{d}^3\mathbf{r}. \end{array} \right\} \qquad (3.14)$$

Thus, $|\Psi(\mathbf{r}, t)|^2$ can be thought of as *probability density for position*. Because of this, the wave function $\Psi(\mathbf{r}, t)$ is often referred to as a *probability amplitude for position.*[2]

---

[2] Note that the Born interpretation of the wave function implies that the position of a particle can be determined precisely at the expense of total uncertainty in its momentum, whereas we showed in Section 1.4 that the minimum uncertainty in the position of a particle of mass m is of the order of h/mc. However, this minimum uncertainty in position is only relevant in relativistic quantum physics.

If we integrate the probability density over all possible positions of the particle, we obtain the probability of finding the particle somewhere in the universe. Because the particle is certain to be found somewhere, this probability must be is equal to unity. Hence the wave function must satisfy the normalization condition

$$\int |\Psi(\mathbf{r}, t)|^2 \, d^3\mathbf{r} = 1, \qquad (3.15)$$

where the integration is over all space.

These equations look less formidable when the particle is restricted to move in one dimension. If the particle moves along the x axis, it may be described by a wave function $\Psi(x, t)$ such that

$$|\Psi(x, t)|^2 \, dx = \left\{ \begin{array}{l} \text{The probability of finding the particle} \\ \text{at time } t \text{ between } x \text{ and } x + dx. \end{array} \right\} \qquad (3.14)$$

Because the particle must be found somewhere between $x = -\infty$ and $x = +\infty$, the wave function must obey the normalization condition

$$\int_{-\infty}^{+\infty} |\Psi(x, t)|^2 \, dx = 1. \qquad (3.17)$$

In practice, a wave function is normalized by multiplying a solution of the Schrödinger equation by an appropriate constant. When we encounter a new wave function we shall often normalize it and then explore the potential positions of the particle by considering the probability density for position. We stress that this exercise only relies on a schematic description of a position measurement, in which a small scale event is amplified and the particle materialises at a specific location with probability proportional to $|\Psi|^2$. This schematic description of a measurement leads to a powerful theory even though no attempt is made to describe how the outcome of the measurement emerges. In fact, quantum mechanics is successful because it avoids explaining how events happen.

3.3   MOMENTUM PROBABILITIES

We shall now explain how the momentum properties of a particle can be described by a Schrödinger wave function. If a wave function can represent a particle with a range of potential positions, it is reasonable to expect that it can also represent a particle with a range of potential momenta. For example, the wave function

$$\Psi(x, t) = A_1 \, e^{+i(k_1 x - \omega_1 t)} + A_2 \, e^{+i(k_2 x - \omega_2 t)},$$

could describe a free particle moving in the x direction with two possible momenta and energies,

$$p_1 = \hbar k_1, E_1 = \hbar \omega_1 \quad \text{and} \quad p_2 = \hbar k_2, E_2 = \hbar \omega_2.$$

This idea is also illustrated by the general solution of the free-particle Schrödinger equation given by Eq. (2.16),

$$\Psi(x, t) = \int_{-\infty}^{+\infty} A(k) e^{i(kx-\omega t)} \, dk, \quad \text{with} \quad \hbar\omega = \frac{\hbar^2 k^2}{2m}. \tag{3.18}$$

In wave terms, we have a superposition of sinusoidal waves each with a different wave number k, and the magnitude of the function $|A(k)|^2$ is a measure of the intensity of the sinusoidal wave with wave number k. In particle terms, this range of wave numbers corresponds to a range of possible momenta $p = \hbar k$. We shall assume, in analogy with the Born interpretation of the wave function, that the most probable momenta found in a measurement correspond to the values of $\hbar k$ for which the function $|A(k)|^2$ is large.

More generally, one may treat position and momentum in a symmetrical way by using Fourier transforms.[3] Any wave function for a particle moving in one dimension can always be written as a Fourier transform

$$\Psi(x, t) = \frac{1}{\sqrt{2\pi\hbar}} \int_{-\infty}^{+\infty} \tilde{\Psi}(p, t) e^{+ipx/\hbar} \, dp. \tag{3.19}$$

The inverse Fourier transform is

$$\tilde{\Psi}(p, t) = \frac{1}{\sqrt{2\pi\hbar}} \int_{-\infty}^{+\infty} \Psi(x, t) e^{-ipx/\hbar} \, dx. \tag{3.20}$$

It can be shown, that if $\Psi(x, t)$ is normalized, then $\tilde{\Psi}(p, t)$ is also normalized; i.e. if

$$\int_{-\infty}^{+\infty} |\Psi(x, t)|^2 \, dx = 1, \quad \text{then} \quad \int_{-\infty}^{+\infty} |\tilde{\Psi}(p, t)|^2 \, dp = 1.$$

This symmetry between position and momentum, and the earlier observation about the interpretation of a superposition of sinusoidal waves, leads us to assume that, since $|\Psi(x, t)|^2$ is the probability density for finding the particle

---

[3] Readers unfamiliar with Fourier transforms should not panic. Eq. (3.19) and Eq. (3.20) are the only Fourier transform equations we shall be using. If you are willing to accept these equations, you can appreciate the symmetry that exists in quantum mechanics between position and momentum observables and also gain an understanding of why these observables can be described by operators.

with position x, $|\widetilde{\Psi}(p, t)|^2$ is the probability density for finding the particle with momentum p. Further, since the wave function $\Psi(x, t)$ is a probability amplitude for position, its Fourier transform $\widetilde{\Psi}(p, t)$ is the probability amplitude for momentum. This generalization of the Born interpretation of the wave function can easily be extended to describe the possible momenta of a particle moving in three dimensions.

3.4 A PARTICLE IN A BOX I

In this section we shall illustrate how position and momentum probability densities can be calculated by considering one of the simplest systems in quantum mechanics: a particle of mass m confined to a one-dimensional region $0 < x < a$. In Section 4.4 we shall show that such a particle has an infinite number of states with discrete energies labelled by a *quantum number* $n = 1, 2, 3, \ldots$. A particle in a state with quantum number n has an energy given by

$$E_n = \frac{\hbar^2 k_n^2}{2m}, \quad \text{where} \quad k_n = \frac{n\pi}{a}, \tag{3.21}$$

and a wave function given by

$$\Psi_n(x, t) = \begin{cases} N \sin k_n x \; e^{-iE_n t/\hbar} & \text{if } 0 < x < a \\ 0 & \text{elsewhere.} \end{cases} \tag{3.22}$$

The constant N, which is called a normalization constant, can be found by normalizing the position probability density. This is given by $|\Psi_n(x, t)|^2$; it is zero outside the region $0 < x < a$, and inside this region it is given by

$$|\Psi_n(x, t)|^2 = N^* \sin k_n x \; e^{+iE_n t/\hbar} \; N \sin k_n x \; e^{-iE_n t/\hbar} = |N|^2 \sin^2 k_n x.$$

The total probability of finding the particle at any of its possible locations is given by the normalization integral,

$$\int_{-\infty}^{+\infty} |\Psi_n(x, t)|^2 \, dx = |N|^2 \int_0^a \sin^2 kx \, dx = |N|^2 \frac{a}{2}.$$

By equating this probability to one, we find that $N = \sqrt{2/a}$ gives rise to a normalized probability density and to a normalized wave function.

The momentum probability density for the particle is given by $|\widetilde{\Psi}_n(p, t)|^2$, where, in accordance with Eq. (3.20),

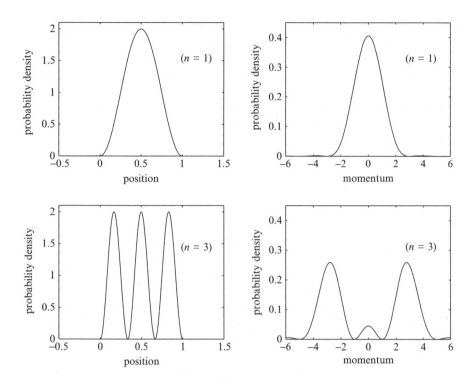

Fig. 3.3 The position and momentum probability densities for a particle confined to the region $0 \le x \le a$ with normalized wave functions given by Eq. (3.22) with $n = 1$ and $n = 3$. In this figure the units are such that a position equal to one corresponds to $x = a$ and a momentum equal to one corresponds to $p = \hbar\pi/a$. Note that the area under each curve is equal to one, signifying that the probability of finding the particle at any of its possible locations or with any of its possible momenta is equal to one.

$$\widetilde{\Psi}_n(p, t) = \frac{1}{\sqrt{2\pi\hbar}} \int_{-\infty}^{+\infty} \Psi_n(x, t)\, e^{-ipx/\hbar}\, dx.$$

If we use Eq. (3.22) with $N = \sqrt{2/a}$, we obtain

$$\widetilde{\Psi}_n(p, t) = \frac{1}{\sqrt{\pi\hbar a}}\, e^{-iE_n t/\hbar} \int_0^a e^{-ipx/\hbar} \sin k_n x\, dx,$$

and the integral can easily be evaluated using

$$\sin k_n x = \frac{e^{+ik_n x} - e^{-ik_n x}}{2i}.$$

In Fig. 3.3 we show position and momentum probability densities for a particle confined to the region $0 < x < a$. Two possible states are considered, the ground state with $n = 1$ and the second excited state with $n = 3$. The position probability densities, shown on the left of the figure, indicate that the most likely location is $x = a/2$ when $n = 1$ and that, when $n = 3$, there are three most likely locations: at $x = a/6$, $x = a/2$ and $x = 5a/6$. The momentum probability densities, shown on the right of the figure, indicate that the most likely momentum is zero when $n = 1$ and that there are two most likely momenta when $n = 3$, at $p = -3\hbar\pi/a$ and at $p = +3\hbar\pi/a$. In fact, a state with a high value for n can be roughly pictured as a particle trapped between $x = 0$ and $x = a$ with two possible momenta, $p = -n\hbar\pi/a$ and $p = +n\hbar\pi/a$.

3.5   EXPECTATION VALUES

In general, the outcome of a measurement in quantum mechanics is a random variable with many possible values. The average of these values is called the expectation value. In principle, the expectation value can be found by taking the average result of measurements on an infinite ensemble of identically prepared systems. Alternatively, we can calculate the expectation value using the probability distribution which governs the outcomes of the measurement. To keep the mathematics as simple as possible we will consider a particle moving along the x axis.

The result of a measurement of the position x is a continuous random variable. The wave function $\Psi(x, t)$ is a probability amplitude for the position observable and $|\Psi(x, t)|^2 \, dx$ is the probability of finding the particle between x and $x + dx$ at time t. Thus, if a measurement of position is repeated many times in an identical way on an identical particle in identical circumstances, many possible outcomes are possible and the expectation value of these outcomes is, according to Eq. (3.6),

$$\langle x \rangle = \int_{+\infty}^{-\infty} x \, |\Psi(x, t)|^2 \, dx. \tag{3.23}$$

Similarly, the measured momentum of the particle is also a continuous random variable. The Fourier transform of the wave function, $\tilde{\Psi}(p, t)$, is the probability amplitude for the momentum observable and $|\tilde{\Psi}(p, t)|^2 dp$ is the probability of a momentum outcome between p and $p + dp$ at time t. Hence, the expectation value for momentum is

$$\langle p \rangle = \int_{-\infty}^{+\infty} p |\tilde{\Psi}(p, t)|^2 \, dp. \tag{3.24}$$

Before proceeding, we shall use $|\Psi|^2 = \Psi^*\Psi$ and $|\tilde{\Psi}|^2 = \tilde{\Psi}^*\tilde{\Psi}$ to rewrite the integrals in Eq. (3.23) and Eq. (3.24) as

$$\langle x \rangle = \int_{-\infty}^{+\infty} \Psi^*(x, t) x\, \Psi(x, t)\, \mathrm{d}x \tag{3.25}$$

and

$$\langle p \rangle = \int_{-\infty}^{+\infty} \tilde{\Psi}^*(p, t) p\, \tilde{\Psi}(p, t)\, \mathrm{d}p. \tag{3.26}$$

These expressions and the Fourier transform integrals, Eq. (3.19) and Eq. (3.20), illustrate the symmetry between position and momentum in quantum physics. However, this symmetry is often hidden because the expectation value for momentum is seldom calculated using $\tilde{\Psi}(p, t)$; it is usually calculated by using the wave function $\Psi(x, t)$ and the expression

$$\langle p \rangle = \int_{-\infty}^{+\infty} \Psi^*(x, t) \left(-i\hbar \frac{\partial}{\partial x} \right) \Psi(x, t)\, \mathrm{d}x. \tag{3.27}$$

This expression, as we shall soon show, can be derived using Fourier transform techniques. You can also show that it is plausible. In problem 9, at the end of the chapter, you are asked to show that Eq. (3.27) predicts that the average values of the uncertain position and momentum of a quantum particle of mass m obey a classical-like relationship:

$$\langle p \rangle = m \frac{\mathrm{d}\langle x \rangle}{\mathrm{d}t}. \tag{3.28}$$

For the benefit of mathematically inclined readers we shall demonstrate that the momentum expectation value given by Eq. (3.27) is identical to that given by Eq. (3.26). We shall do this by using the relations between $\Psi(x, t)$ and $\tilde{\Psi}(p, t)$ given by Eq. (3.19) and Eq. (3.20).

Using Eq. (3.19) we write

$$\left(-i\hbar \frac{\partial}{\partial x} \right) \Psi(x, t) = \left(-i\hbar \frac{\partial}{\partial x} \right) \left(\frac{1}{\sqrt{2\pi\hbar}} \int_{-\infty}^{+\infty} \tilde{\Psi}(p, t)\, \mathrm{e}^{+ipx/\hbar}\, \mathrm{d}p \right)$$

which gives

$$\left(-i\hbar \frac{\partial}{\partial x} \right) \Psi(x, t) = \left(\frac{1}{\sqrt{2\pi\hbar}} \int_{-\infty}^{+\infty} p\tilde{\Psi}(p, t)\, \mathrm{e}^{+ipx/\hbar}\, \mathrm{d}p \right).$$

We can now rewrite Eq. (3.27) as

$$\langle p \rangle = \int_{-\infty}^{+\infty} \Psi^*(x, t) \left(\frac{1}{\sqrt{2\pi\hbar}} \int_{-\infty}^{+\infty} p\tilde{\Psi}(p, t)\, e^{+ipx/\hbar}\, \mathrm{d}p \right) \mathrm{d}x.$$

If the order of integration is interchanged, we obtain

$$\langle p \rangle = \int_{-\infty}^{+\infty} \left(\frac{1}{\sqrt{2\pi\hbar}} \int_{-\infty}^{+\infty} \Psi^*(x, t)\, e^{+ipx/\hbar}\, \mathrm{d}x \right) p\tilde{\Psi}(p, t)\, \mathrm{d}p.$$

We now use Eq. (3.20) to show that the integral in the brackets is equal to $\tilde{\Psi}^*(p, t)$ and obtain

$$\langle p \rangle = \int_{-\infty}^{+\infty} \tilde{\Psi}^*(p, t)\, p\, \tilde{\Psi}(p, t)\, \mathrm{d}p$$

which is identical to Eq. (3.26).

Operators

We shall now introduce an idea which will become increasingly important as we develop the basic elements of quantum mechanics. This is the idea that observables in quantum mechanics can be described by operators. At this stage we shall consider the role of operators in the calculation of position and momentum expectation values using Eq. (3.25) and Eq. (3.27). The recipe for the calculation is as follows:

- First prepare a sandwich with Ψ^* and Ψ.

- To find $\langle x \rangle$ insert x into the sandwich, and to find $\langle p \rangle$ insert $-i\hbar\, \partial/\partial x$ into the sandwich.

- Then integrate over x.

In this recipe, the position observable is represented by x and the momentum observable is represented by $-i\hbar\, \partial/\partial x$. However, both x and $-i\hbar\, \partial/\partial x$ can be considered as operators which act on the wave function; the real number x merely multiplies $\Psi(x, t)$ by a factor x, and the differential expression $-i\hbar\, \partial/\partial x$ differentiates $\Psi(x, t)$ and multiplies it by $-i\hbar$. To emphasize the role of operators in quantum physics we rewrite Eq. (3.25) and Eq. (3.27) as

$$\langle x \rangle = \int_{-\infty}^{+\infty} \Psi^*(x, t)\, \hat{x}\, \Psi(x, t)\, \mathrm{d}x \quad \text{and} \quad \langle p \rangle = \int_{-\infty}^{+\infty} \Psi^*(x, t)\, \hat{p}\, \Psi(x, t)\, \mathrm{d}x.$$

$$(3.29)$$

The circumflex accent denotes an operator, and the operators for the position and momentum observables are

$$\hat{x} = x \quad \text{and} \quad \hat{p} = -i\hbar \frac{\partial}{\partial x}. \tag{3.30}$$

The generalization of these ideas to a particle moving in three dimensions is straightforward: the operators for position and momentum are

$$\hat{\mathbf{r}} = \mathbf{r} \quad \text{and} \quad \hat{\mathbf{p}} = -i\hbar \nabla \tag{3.31}$$

and the expectation values for position and momentum are given by the three-dimensional sandwich integrals,

$$\langle \mathbf{r} \rangle = \int \Psi^*(\mathbf{r}, t) \, \hat{\mathbf{r}} \, \Psi(\mathbf{r}, t) \, \mathrm{d}^3 r \quad \text{and} \quad \langle \mathbf{p} \rangle = \int \Psi^*(\mathbf{r}, t) \, \hat{\mathbf{p}} \, \Psi(\mathbf{r}, t) \, \mathrm{d}^3 r. \tag{3.32}$$

Finally, we emphasise that the order in which functions and operators are written down is important; for example,

$$\Psi^*(\mathbf{r}, t) \, \hat{\mathbf{p}} \, \Psi(\mathbf{r}, t) \neq \hat{\mathbf{p}} \, \Psi^*(\mathbf{r}, t) \, \Psi(\mathbf{r}, t).$$

Uncertainties

Operators can also be used to calculate uncertainties in position and momentum. According to Eq. (3.8) these uncertainties are given by

$$\Delta x = \sqrt{\langle x^2 \rangle - \langle x \rangle^2} \quad \text{and} \quad \Delta p = \sqrt{\langle p^2 \rangle - \langle p \rangle^2}. \tag{3.33}$$

The expectation values of x and p are given by Eq. (3.29), and the expectation values of x^2 and p^2 can be found from the generalization of this equation. They are given by the sandwich integrals

$$\langle x^2 \rangle = \int_{-\infty}^{+\infty} \Psi^*(x, t) \, \hat{x}^2 \, \Psi(x, t) \, \mathrm{d}x$$

$$\text{and} \quad \langle p^2 \rangle = \int_{-\infty}^{+\infty} \Psi^*(x, t) \, \hat{p}^2 \, \Psi(x, t) \, \mathrm{d}x, \tag{3.34}$$

where the operator $\hat{x}^2$ is equivalent to $\hat{x} \times \hat{x}$ and $\hat{p}^2$ is equivalent to $\hat{p} \times \hat{p}$. Using Eq. (3.30), we have

$$\hat{x}^2 = x^2 \quad \text{and} \quad \hat{p}^2 = -\hbar^2 \frac{\partial^2}{\partial x^2}. \tag{3.35}$$

The actual values for the uncertainties in position and momentum depend upon the form of the wave function. For example, in problem 4 at the end of this chapter, you are asked to show that the uncertainties in the position and momentum for a particle with the wave function

$$\Psi(x) = N\,e^{-x/2a^2}$$

are $\Delta x = a/\sqrt{2}$ and $\Delta p = \hbar/a\sqrt{2}$. We note that, in this example, the product of the uncertainties obey the relation

$$\Delta x\,\Delta p = \frac{\hbar}{2},$$

which is in accordance with the general statement of the Heisenberg uncertainty principle given by Eq. (1.15). For all other wave functions the product $\Delta x\,\Delta p$ is larger, as illustrated in problems 5, 6 and 7.

3.6   QUANTUM STATES

We began this chapter with the puzzle of how a wave function can describe both the wave-like and the particle-like properties of a quantum particle. We have seen that this puzzle can be resolved if the wave function of a particle governs the potential outcomes of measurements on the particle. This interpretation of the wave function has transformed the philosophical basis of physics. Physics no longer tries to predict exactly what will happen; it is now content with predicting the probabilities of a range of possible outcomes.

It is not clear whether probability is being used in quantum mechanics, as in the kinetic theory of gases, to cover up our ignorance of some underlying specific description of the particle, a description which assigns an exact position to the particle which is revealed by the measurement, or whether probability is being used to provide a complete and fundamental description of the particle. In the latter case, it is pointless or meaningless to speculate on where the particle is prior to the measurement. Its exact position is not revealed by the measurement, but brought into existence by the measurement; the particle, like the experimenter, is surprised by the outcome!

However, these issues, which have been debated since the inception of quantum mechanics, are not our immediate concern.[4] Our main aim is to appreciate the power and elegance of quantum mechanics as a consistent theory of microscopic phenomena and we shall do this by developing an intuitive understanding of wave functions.

---

[4] An excellent account of these issues and an extensive list of references are provided by F. Lalö, *American Journal of Physics*, vol. 69, page 655 (2001).

The key idea is that a wave function represents a *quantum state* of a particle, a state of motion that bears only a passing resemblance to the well-defined trajectories of classical physics.[5] In the first three chapters of this book we have touched upon the general properties of quantum states and these properties will become more understandable as we consider examples of quantum states in the chapters that follow. At this stage, it is useful to set out the following important properties of quantum states:

- In the absence of measurements, a quantum state evolves with time smoothly and deterministically in accordance with the time-dependent Schrödinger equation, Eq. (2.17).

- A quantum state describes potentialities which can become realities. As illustrated in this chapter, a quantum state can predict the possible outcomes of position and momentum measurements and the probabilities for the occurrence of these outcomes. More generally, a quantum state can predict the possible outcomes of any measurement.

- A quantum state is a linear superposition of other quantum states which means that a particle in one quantum state is also simultaneously in other quantum states. This property is called the *principle of linear superposition*. We used this principle when we wrote down the wave function for a particle passing through two slits, Eq. (3.11), and when we argued in Section 3.3 that a wave function can describe a particle with a range of possible momenta. It will be used again in Chapter 4 when we consider a wave function of a particle with a range of possible energies.

- Lastly, a quantum state is fragile. When a measurement occurs, a quantum state is destroyed and replaced by a new quantum state which is compatible with the random outcome of the measurement. This abrupt and non-deterministic process is called *the collapse of the wave function*. We used this idea to explain why a two-slit experiment, in which the slit through which each particle passes is identified, does not give rise to an interference pattern. To a considerable extent, the collapse of the wave function is an arbitrary rule that bridges the gap between the unobserved quantum system and the observed quantum system. The underlying mechanism for the collapse is not understood.

---

[5] In the most general formulation of quantum mechanics, quantum states are complex vectors. Some quantum states can also be represented by wave functions and others, like quantum states describing the spin properties of an electron, by matrices.

PROBLEMS 3

1. This problem considers the Poisson distribution, a probability distribution for a discrete random variable which was first used by Siméon-Denis Poisson to describe seemingly random criminal events in Paris in 1837. If independent events have a constant tendency to occur and if the average rate of occurrence is a, then the probability that n events actually occur is given by

$$p_n = \frac{e^{-a}a^n}{n!} \quad \text{with} \quad n = 0, 1, 2, \ldots \infty.$$

(a) By noting that

$$e^{+a} = 1 + \frac{a}{1!} + \frac{a^2}{2!} + \frac{a^3}{3!} + \cdots$$

show that

$$\sum_{n=0}^{n=\infty} p_n = 1,$$

thereby verifying that the Poisson distribution is normalized.

(b) By using $n/n! = 1/(n-1)!$ and $a^n = aa^{n-1}$, show that

$$\sum_{n=0}^{n=\infty} np_n = a,$$

thereby verifying that the average rate of occurrence, or the expectation value $\langle n \rangle$, is equal to a.

(c) By using similar techniques, find $\langle n^2 \rangle$ and show, using Eq. (3.4), that the standard deviation of the Poisson distribution is given by

$$\Delta n = \sqrt{a}.$$

2. This problem considers a Gaussian distribution, with standard deviation σ, given by

$$\rho(x)dx = \frac{1}{\sqrt{2\pi\sigma^2}} e^{-x^2/2\sigma^2} dx \quad \text{with} \quad -\infty < x < +\infty.$$

This probability distribution is normalized because

$$\frac{1}{\sqrt{2\pi\sigma^2}} \int_{-\infty}^{+\infty} e^{-x^2/2\sigma^2} \, dx = 1.$$

It is famous for describing a random variable which arises from a multitude of small random contributions, such as the net distance travelled by a tottering drunk with very small legs.

(a) By considering the effect of the transformation $x \to -x$ on the function $x\rho(x)$, show that the expectation value of x is equal to zero.

(b) By using

$$\int_{-\infty}^{+\infty} \rho(x) \, dx = \left[x \, \rho(x) \right]_{-\infty}^{+\infty} - \int_{-\infty}^{\infty} x \, \frac{d\rho}{dx} \, dx,$$

show that the expectation value of x^2 is equal to σ^2.

(c) Hence verify that the standard deviation of x is equal to σ.

3. The lifetime of an unstable particle is governed by the exponential probability distribution. In particular, the probability that the particle lives for time t and then decays in the time interval t to $t + dt$ is given by

$$p(t) \, dt = e^{-\lambda t} \, \lambda \, dt,$$

where λ is a positive decay constant.

(a) Show that the probability that the particle eventually decays is equal to one.

(b) Find an expression for the mean lifetime of the particle.

(c) Find an expression for the probability that the particle lives for at least time T.

4. In this question you should avoid unnecessary mathematics by using the properties of the Gaussian probability distribution given in problem 2.
    Consider a particle of mass m described by the wave function

$$\Psi(x) = N e^{-x^2/2a^2}$$

where a is a constant length.

(a) Use the properties of the Gaussian probability distribution to confirm that the expectation values of the position and the square of the position are

$$\langle x \rangle = 0 \quad \text{and} \quad \langle x^2 \rangle = \frac{a^2}{2}.$$

(b) Show, without lengthy calculation, that the expectation values of the momentum and the square of the momentum are

$$\langle p \rangle = 0 \quad \text{and} \quad \langle p^2 \rangle = \frac{\hbar^2}{2a^2}.$$

[Hint: I suggest you use your skill at integration by parts to show that

$$\int_{-\infty}^{+\infty} \Psi^* \frac{d^2 \Psi}{dx^2} \, dx = \int_{-\infty}^{+\infty} \frac{d\Psi^*}{dx} \frac{d\Psi}{dx} \, dx,$$

and also make use of the integrals used in part (a).]

(c) Hence show that the uncertainty in position, Δx, and the uncertainty in momentum, Δp, for this particle obey the relation

$$\Delta x \, \Delta p = \frac{\hbar}{2}.$$

5. A particle is confined to a region $0 \leq x \leq a$ and has a wave function of the form

$$\psi(x) = Nx(a - x),$$

where N is a constant.

(a) Normalize the wave function and find the average position of the particle.

(b) Show that the uncertainties in the position and momentum of the particle are given by

$$\Delta x = \sqrt{\frac{1}{28}} \, a \quad \text{and} \quad \Delta p = \sqrt{10} \, \frac{\hbar}{a}.$$

6. A particle of mass m is confined to a region $-a/2 < x < +a/2$. Outside this region the wave function is zero and inside it is given by

$$\Psi = \sqrt{\frac{2}{a}} \cos \frac{\pi x}{a}.$$

(a) Explain the physical significance of the integrals

$$\int_{-\infty}^{+\infty} \Psi^* x \Psi \, dx$$

and

$$\int_{-\infty}^{+\infty} \Psi^* \left(-i\hbar \frac{d}{dx} \right) \Psi \, dx.$$

Show that both integrals are zero.

(b) Show that Δx, the uncertainty in the position of the particle, and Δp, the uncertainty in the momentum of the particle, are related by

$$\Delta x \, \Delta p = \sqrt{\frac{\pi^2 - 6}{12}} \, \hbar.$$

(The following integral

$$\int_{-a/2}^{+a/2} x^2 \cos^2 \frac{\pi x}{a} \, dx = \frac{a^3}{\pi^3} \left(\frac{\pi^3}{24} - \frac{\pi}{4} \right)$$

is useful.)

7. Consider a particle with normalized wave function

$$\Psi(x) = \begin{cases} Nx \, e^{-\alpha x/2} & \text{if } 0 \le x < \infty \\ 0 & \text{elsewhere,} \end{cases}$$

where α is a positive real constant and $N = \sqrt{\alpha^3/2}$.

(a) Write down an expression for the probability of finding the particle between x and $x + dx$. Illustrate how this probability depends on x and find the most probable value of x.

(b) Find the expectation values for the position and the square of the position, $\langle x \rangle$ and $\langle x^2 \rangle$.

(c) Find the expectation values for the momentum and the square of the momentum, $\langle p \rangle$ and $\langle p^2 \rangle$.

(d) Show that these expectation values yield uncertainties for position and momentum which are consistent with the Heisenberg uncertainty relation.

(The mathematical identity

$$\int_0^\infty e^{-\alpha x} x^n \, dx = \frac{n!}{\alpha^{n+1}} \quad \text{for} \quad n > -1$$

is useful.)

8. In this problem the probability density for the position of a particle with wave function $\Psi(x, t)$ is denoted by $\rho(x, t)$. In general, the value of $\rho(x, t)$ in a particular region will change with time and this change can be attributed to the flow of probability into and out of the region. Indeed we expect that there is a probability current density $j(x, t)$ which obeys the continuity equation[6]

$$\frac{\partial \rho}{\partial t} = -\frac{\partial j}{\partial x}.$$

When there are many particles, the actual current of particles is obtained by multiplying $j(x, t)$ by the actual density of particles.

(a) By noting that the time dependence of the wave function is governed by the Schrödinger equation,

$$i\hbar \frac{\partial \Psi}{\partial t} = \left[-\frac{\hbar^2}{2m} \frac{\partial^2}{\partial x^2} + V(x) \right] \Psi,$$

---

[6] The reader may have come across a similar equation when considering the conservation of charge in electromagnetism or particles in diffusion. In three-dimensional problems it has the form $\partial \rho / \partial t = -\text{div} \mathbf{j}$. It is a book-keeping equation which describes the flow of probability. Its book-keeping properties can be illustrated by considering a one-dimensional region between $x = x_1$ and $x = x_2$. The amount of probability in this region, i.e. the probability of finding the particle in the region at time t is

$$\int_{x_1}^{x_2} \rho(x, t) \, dx.$$

The rate of change of probability in this region is given by

$$\frac{d}{dt}\left[\int_{x_1}^{x_2} \rho \, dx \right] = \int_{x_1}^{x_2} \frac{\partial \rho}{\partial t} \, dx = -\int_{x_1}^{x_2} \frac{\partial j}{\partial x} \, dx = j(x_1, t) - j(x_2, t).$$

We see that the rate of change of probability inside the region is the difference between the current into the region at $x = x_1$ and the current out at $x = x_2$.

derive an expression for the time derivative of

$$\rho(x, t) = \Psi^*(x, t)\Psi(x, t).$$

(b) Hence show that the probability current is given by

$$j(x, t) = \frac{i\hbar}{2m}\left[\Psi^*\frac{\partial\Psi}{\partial x} - \Psi\frac{\partial\Psi^*}{\partial x}\right].$$

9. In this problem you are asked to use Eqs. (3.25) and (3.27) to show that the expectation values of the position and momentum of a particle with mass m are related by

$$m\frac{\mathrm{d}\langle x\rangle}{\mathrm{d}t} = \langle p\rangle.$$

The method is similar to that used in the previous problem.

(a) By noting that the time dependence of the wave function is governed by the Schrödinger equation, show that

$$\frac{\mathrm{d}(\Psi^*x\Psi)}{\mathrm{d}t} = \frac{i\hbar}{2m}\left[x\Psi^*\frac{\partial^2\Psi}{\partial x^2} - x\Psi\frac{\partial^2\Psi^*}{\partial x^2}\right],$$

and show that this can be rewritten as

$$\frac{\mathrm{d}(\Psi^*x\Psi)}{\mathrm{d}t} = \frac{i\hbar}{2m}\frac{\partial}{\partial x}\left[x\Psi^*\frac{\partial\Psi}{\partial x} - x\Psi\frac{\partial\Psi^*}{\partial x}\right] - \frac{i\hbar}{2m}\left[\Psi^*\frac{\partial\Psi}{\partial x} - \Psi\frac{\partial\Psi^*}{\partial x}\right].$$

(b) Assuming that the wave function tends to zero sufficiently rapidly at $x = \pm\infty$, show that

$$\frac{\mathrm{d}}{\mathrm{d}t}\int_{-\infty}^{+\infty}\Psi^*x\Psi\,\mathrm{d}x = -\frac{i\hbar}{2m}\int_{-\infty}^{+\infty}\left[\Psi^*\frac{\partial\Psi}{\partial x} - \Psi\frac{\partial\Psi^*}{\partial x}\right]\mathrm{d}x.$$

(c) Now integrate by parts and show that

$$m\frac{\mathrm{d}}{\mathrm{d}t}\int_{-\infty}^{+\infty}\Psi^*x\Psi\,\mathrm{d}x = \int_{-\infty}^{+\infty}\Psi^*(x, t)\left(-i\hbar\frac{\partial}{\partial x}\right)\Psi(x, t)\,\mathrm{d}x.$$

4

Energy and time

This chapter focuses on how the energy observable is described in quantum mechanics and how it is related to the time evolution of a quantum state. Most importantly, we shall show that the observable properties of a quantum state with a sharply defined energy never change.

4.1 THE HAMILTONIAN OPERATOR

In quantum mechanics observable quantities are described by operators. In the last chapter we found that the position and momentum observables are described by

$$\hat{\mathbf{r}} = \mathbf{r} \quad \text{and} \quad \hat{\mathbf{p}} = -i\hbar\nabla. \tag{4.1}$$

The energy observable in quantum mechanics is described by an operator called the Hamiltonian operator and it is denoted by $\hat{H}$. We shall assume that the relation between the operators for energy, momentum and position is similar to the relation between the classical energy, momentum and position. In particular, we shall assume that the Hamiltonian operator for a particle with mass m in a potential energy field $V(\mathbf{r})$ is given by

$$\hat{H} = \frac{\hat{\mathbf{p}}^2}{2m} + V(\hat{\mathbf{r}}). \tag{4.2}$$

We can rewrite this using Eq. (4.1) to give

$$\hat{H} = -\frac{\hbar^2}{2m}\nabla^2 + V(\mathbf{r}). \tag{4.3}$$

We emphasize that the Hamiltonian operator $\hat{H}$ has a dual role in quantum mechanics. First, the operator $\hat{H}$ describes the energy observable; for example, the recipe for expectation values given in Section 3.4 implies that the energy expectation value at time t for a particle with wave function $\Psi(\mathbf{r}, t)$ is

$$\langle E \rangle = \int \Psi^*(\mathbf{r}, t)\, \hat{H}\, \Psi(\mathbf{r}, t)\, \mathrm{d}^3\mathbf{r}. \tag{4.4}$$

Second, the operator $\hat{H}$ governs the time evolution of the wave function because the Schrödinger equation, Eq. (2.17), has the form

$$i\hbar \frac{\partial \Psi}{\partial t} = \hat{H}\Psi. \tag{4.5}$$

Thus, in quantum mechanics, there is a fundamental connection between energy and time.

We shall explore this connection by finding solutions to the Schrödinger equation. The procedure used will be identical to that used to solve the classical wave equation or the diffusion equation: we shall seek a *separable solution* and then solve an *eigenvalue problem*. Because this procedure seems arbitrary in the abstract, it is best introduced by considering a familiar and simple problem, the problem of finding the normal modes of a vibrating string.

4.2   NORMAL MODES OF A STRING

Let $\Psi(x, t)$ represent the transverse displacement of a stretched string at the point x at time t. This displacement is governed by the classical wave equation

$$\frac{\partial^2 \Psi}{\partial x^2} - \frac{1}{c^2} \frac{\partial^2 \Psi}{\partial t^2} = 0, \tag{4.6}$$

where c is the speed of waves on the string. If the ends of the string are fixed at $x = 0$ and $x = a$, we seek solutions to the wave equation which satisfy the boundary conditions

$$\Psi(x, t) = 0 \text{ at } x = 0 \text{ and at } x = a, \quad \text{for all time } t. \tag{4.7}$$

There are an infinite number of such solutions because the string can vibrate in an infinite number of ways.

The normal mode solutions are particularly simple. They correspond to vibrations where all points on the string move with the same time dependence. They have the separable form

$$\Psi(x, t) = \psi(x)T(t).$$

$$(4.8)$$

The function $T(t)$ describes the common time dependence of each point and the function $\psi(x)$ describes the spatial shape of the vibration. If we substitute the separable form for $\Psi(x, t)$ into the wave equation, Eq. (4.6), and if we carefully separate out the functions which depend on t from those which depend on x we obtain

$$\frac{1}{T}\frac{d^2 T}{dt^2} = \frac{c^2}{\psi}\frac{d^2 \psi}{dx^2}.$$

$$(4.9)$$

The equal sign in this equation asserts that a function of t on the left-hand side is equal to a function of x on the right-hand side, for all x and t. This can be true only if both functions are equal to the same constant. We shall denote this constant by $-\omega^2$ and we shall also set $\omega = ck$ where k is another constant. By equating both the left-hand side and the right-hand side of Eq. (4.9) to the constant $-\omega^2$, we find the time dependence $T(t)$ and spatial shape $\psi(x)$ for each normal mode solution.

The time dependence $T(t)$ is governed by the differential equation

$$\frac{d^2 T}{dt^2} = -\omega^2 T.$$

$$(4.10)$$

The general solution is

$$T(t) = A \cos \omega t + B \sin \omega t,$$

$$(4.11)$$

where A and B are arbitrary constants. This solution describes sinusoidal motion with an angular frequency ω, which, as yet, is undetermined.

The normal mode function $\psi(x)$ is governed by the differential equation

$$\frac{d^2 \psi}{dx^2} = -k^2 \psi, \quad \text{where} \quad k = \frac{\omega}{c}.$$

$$(4.12)$$

And it also satisfies the boundary conditions

$$\psi(0) = \psi(a) = 0,$$

$$(4.13)$$

which follow from Eqs. (4.7) and (4.8). The general solution of the differential equation (4.12) is

$$\psi(x) = M \cos kx + N \sin kx,$$

where M and N are arbitrary constants. The boundary condition at $x = 0$ gives $M = 0$ and the boundary condition at $x = a$ restricts the values of k to

$\pi/a, 2\pi/a, 3\pi/a$, etc. Thus, there are an infinite number of normal mode solutions with spatial shapes given by

$$\psi_n(x) = N \sin k_n x, \quad \text{with} \quad k_n = \frac{n\pi}{a}, \tag{4.14}$$

where $n = 1, 2, 3, \ldots$. The spatial shapes of the normal modes of a string with $n = 1, 2, 3$, and 4 are shown in Fig. 4.1.

If these spatial functions are combined with the time-dependent functions, Eq. (4.11), with angular frequencies $\omega_n = ck_n$, we obtain a complete specification of the normal mode solutions. They have the form

$$\Psi_n(x, t) = [A_n \cos \omega_n t + B_n \sin \omega_n t] \sin k_n x. \tag{4.15}$$

Because the classical wave equation, Eq. (4.6), is a homogeneous linear partial differential equation, a linear superposition of normal mode solutions

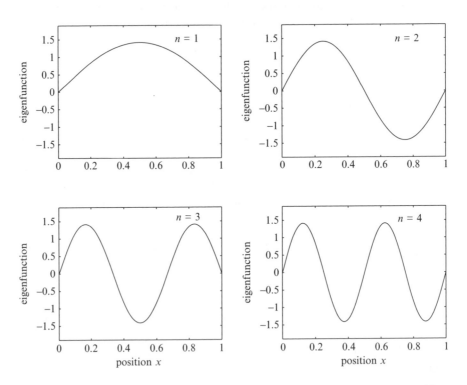

Fig. 4.1 The spatial shapes of four eigenfunctions $\psi_n(x)$ of the eigenvalue problem defined by the differential equation (4.12) and the boundary conditions (4.13) with $a = 1$. In classical physics, these eigenfunctions may describe the shape of a normal mode of vibration of a string with definite angular frequency. In quantum physics, they may describe the shape of a wave function of a particle in a box with definite energy.

is also a solution. Indeed, it can be shown that the general motion of a vibrating string with fixed ends is given by

$$\Psi(x, t) = \sum_{n=1, 2, 3...} [A_n \cos \omega_n t + B_n \sin \omega_n t] \sin k_n x. \qquad (4.16)$$

If the initial displacement and velocity of each point of the string are known, Fourier series techniques can be used to find the constants, A_n and B_n, for each term in the series.

It is useful to describe what we have done using the mathematical language that is used in quantum mechanics. In finding the normal mode solutions of a vibrating string we have solved an eigenvalue problem. The eigenvalue problem is defined by the differential equation (4.12) and the boundary conditions (4.13). We note that the differential equation contains an undetermined parameter k and that solutions to the eigenvalue problem only exist for particular values of k given in Eq. (4.14). The function $\psi_n(x) = N \sin k_n x$ is called the eigenfunction belonging to the eigenvalue $k_n = n\pi/a$. Once we found the eigenfunctions, we wrote down the general motion in terms of a linear superposition of eigenfunctions, Eq. (4.16).

4.3   STATES OF CERTAIN ENERGY

We shall now see how the mathematics of a vibrating string may be adapted to find the solutions of the Schrödinger equation. We shall begin by showing how solutions analogous to normal mode solutions may be constructed.

According to Eq. (2.17), the Schrödinger equation for a particle of mass m in potential $V(\mathbf{r})$ is

$$i\hbar \frac{\partial \Psi}{\partial t} = \left[-\frac{\hbar^2}{2m} \nabla^2 + V(\mathbf{r}) \right] \Psi. \qquad (4.17)$$

As in the classical example of a vibrating string, we shall seek separable solutions of the form

$$\Psi(\mathbf{r}, t) = \psi(\mathbf{r})T(t). \qquad (4.18)$$

If we substitute into the Schrödinger equation and carefully separate functions depending on t from those which depend on $\mathbf{r}$ we obtain

$$\frac{i\hbar}{T}\frac{dT}{dt} = \frac{1}{\psi}\left[-\frac{\hbar^2}{2m} \nabla^2 \psi + V(\mathbf{r})\psi \right]. \qquad (4.19)$$

The equal sign in this equation asserts that a function of t on the left-hand side is equal to a function of $\mathbf{r}$ on the right-hand side, for all $\mathbf{r}$ and t. This can be true only if both functions are equal to the same constant. We shall denote this constant by E. By equating both the left-hand side and the right-hand side of Eq. (4.19) to the constant E, we can find the time dependence $T(t)$ and spatial shape $\psi(\mathbf{r})$ for each separable solution of the Schrödinger equation.

The time dependence $T(t)$ is governed by the differential equation

$$\frac{dT}{dt} = \left(\frac{-iE}{\hbar}\right)T. \tag{4.20}$$

The general solution is

$$T(t) = A\,e^{-iEt/\hbar}, \tag{4.21}$$

where A is an arbitrary constant. The constant E is, as yet, undetermined, but some readers may have guessed its meaning.

The spatial shape $\psi(\mathbf{r})$ of the wave function is governed by the differential equation

$$\left[-\frac{\hbar^2}{2m}\nabla^2 + V(\mathbf{r})\right]\psi(\mathbf{r}) = E\psi(\mathbf{r}), \tag{4.22}$$

which may be rewritten more succinctly as

$$\hat{H}\psi(\mathbf{r}) = E\psi(\mathbf{r}), \tag{4.23}$$

where $\hat{H}$ is the Hamiltonian operator. This equation is called *the energy eigenvalue equation* and the function $\psi(\mathbf{r})$ is called *the eigenfunction of $\hat{H}$ belonging to the eigenvalue E*; in practice, there are many eigenvalues and many eigenfunctions. Equation (4.23) is also called *the time-independent Schrödinger equation*.

An eigenfunction of the Hamiltonian operator is a very special mathematical function. When the complicated operator $\hat{H}$ acts upon a function we expect a mess, but when it operates on an eigenfunction it gives the *same* function multiplied by a constant, as shown in Eq. (4.23). If the eigenfunction belonging to the eigenvalue E is combined with the time-dependent function (4.21), we obtain a special solution to the Schrödinger equation:

$$\Psi(\mathbf{r}, t) = \psi(\mathbf{r})\,e^{-iEt/\hbar}. \tag{4.24}$$

We shall now show that this wave function represents a state with a sharply defined energy E.

In general, when the energy is measured, the outcome is uncertain. In analogy with Eq. (3.33), which gives the uncertainties for position and momentum, the uncertainty in the energy is given by

$$\Delta E = \sqrt{\langle E^2 \rangle - \langle E \rangle^2}, \tag{4.25}$$

where $\langle E \rangle$ is the expectation value for the energy and $\langle E^2 \rangle$ is the expectation value for the square of the energy. These expectation values, for a particle with a normalized wave function $\Psi(\mathbf{r}, t)$, are given by the sandwich integrals

$$\langle E \rangle = \int \Psi^*(\mathbf{r}, t) \hat{H} \Psi(\mathbf{r}, t) \, d^3\mathbf{r} \tag{4.26}$$

and

$$\langle E^2 \rangle = \int \Psi^*(\mathbf{r}, t) \hat{H}^2 \Psi(\mathbf{r}, t) \, d^3\mathbf{r}. \tag{4.27}$$

It is easy to evaluate these integrals when the wave function is given by Eq. (4.24). In this special case, the wave function $\Psi(\mathbf{r}, t)$, like $\psi(\mathbf{r})$, is an eigenfunction of the Hamiltonian operator with eigenvalue E and we can use

$$\hat{H} \Psi(\mathbf{r}, t) = E \Psi(\mathbf{r}, t) \tag{4.28}$$

to give

$$\langle E \rangle = \int \Psi^*(\mathbf{r}, t) \hat{H} \Psi(\mathbf{r}, t) \, d^3\mathbf{r} = E \int \Psi^*(\mathbf{r}, t) \Psi(\mathbf{r}, t) \, d^3\mathbf{r} = E.$$

Moreover, if Ψ is an eigenfunction of $\hat{H}$, it is also an eigenfunction of the product of $\hat{H}$ with $\hat{H}$. Using

$$\hat{H}^2 \Psi(\mathbf{r}, t) = E^2 \Psi(\mathbf{r}, t), \tag{4.29}$$

we obtain

$$\langle E^2 \rangle = \int \Psi^*(\mathbf{r}, t) \hat{H}^2 \Psi(\mathbf{r}, t) \, d^3\mathbf{r} = E^2 \int \Psi^*(\mathbf{r}, t) \Psi(\mathbf{r}, t) \, d^3\mathbf{r} = E^2.$$

When we substitute $\langle E \rangle = E$ and $\langle E^2 \rangle = E^2$ into Eq. (4.25), we find that the uncertainty in energy ΔE is zero. These results imply that the result of an energy measurement is certain to be E when the wave function is an eigenfunction of the Hamiltonian with eigenvalue E. We conclude that an eigenfunction of the Hamiltonian always describes a state of definite energy.

4.4   A PARTICLE IN A BOX II

One of the key features of quantum physics is that the possible energies of a confined particle are quantized. Indeed, the familiar quantized energy levels of atomic, nuclear and particle physics are manifestations of confinement. We shall illustrate the connection between confinement and quantized energy levels by considering a particle confined to a box.

A one-dimensional box

We begin by considering a particle moving in one dimension with potential energy

$$V(x) = \begin{cases} 0 & \text{if } 0 < x < a \\ \infty & \text{elsewhere,} \end{cases} \tag{4.30}$$

This infinite square-well potential confines the particle to a one-dimensional box of size a, as shown in Fig. 4.2. In classical physics, the particle either lies at the bottom of the well with zero energy or it bounces back and forth between the barriers at $x = 0$ and $x = a$ with any energy up to infinity. In quantum physics, more varied states exist. Each is described by a wave function $\Psi(x, t)$ which obeys the one-dimensional Schrödinger equation

$$i\hbar \frac{\partial \Psi}{\partial t} = \left[-\frac{\hbar^2}{2m} \frac{\partial^2}{\partial x^2} + V(x) \right] \Psi. \tag{4.31}$$

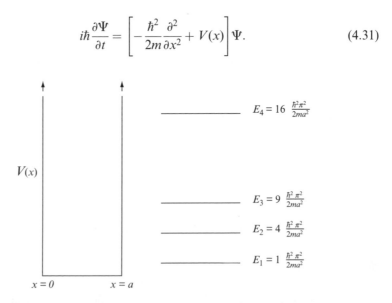

Fig. 4.2   Low-lying energy levels of a particle of mass m confined by an infinite square-well potential $V(x)$ with width a.

However, when the particle has a definite value E, the wave function has the special form

$$\Psi(x, t) = \psi(x)\,e^{-iEt/\hbar},\tag{4.32}$$

where $\psi(x)$ satisfies the energy eigenvalue equation

$$\left[-\frac{\hbar^2}{2m}\frac{d^2}{dx^2} + V(x)\right]\psi(x) = E\psi(x).\tag{4.33}$$

We shall now seek physically acceptable solutions to Eq. (4.33). Because the potential energy $V(x)$ rises abruptly to infinity at $x = 0$ and at $x = a$, the particle is confined to the region $0 < x < a$ and outside this region the eigenfunction $\psi(x)$ is zero. Inside this region, the potential energy is zero and the eigenfunction is a solution of Eq. (4.33) with $V(x) = 0$. We shall simplify this equation by rewriting the energy E as

$$E = \frac{\hbar^2 k^2}{2m}\tag{4.34}$$

to give

$$\frac{d^2\psi}{dx^2} = -k^2\psi.\tag{4.35}$$

Physically acceptable solutions of this differential equation are obtained by writing down the general solution

$$\psi(x) = M\cos kx + N\sin kx,$$

where M and N are constants, and by imposing boundary conditions

$$\psi(0) = \psi(a) = 0,\tag{4.36}$$

which ensure that the position probability density of the particle does not change abruptly at the edge of the box.

Readers should note that the energy eigenvalue problem for a particle in a one-dimensional box, defined by the differential equation (4.35) and boundary conditions (4.36), is identical to the eigenvalue problem for a vibrating string defined by Eqs. (4.12) and (4.13). In both cases, there are an infinite number of eigenfunctions labelled by an integer $n = 1, 2, 3, \ldots$. They are given by

$$\psi_n(x) = N\sin k_n x, \quad \text{with} \quad k_n = \frac{n\pi}{a},\tag{4.37}$$

where N is an arbitrary constant, and they are illustrated in Fig. 4.1. In classical physics, the eigenfunctions ψ_n can be used to describe the possible shapes of normal modes of vibration of a string. In quantum physics, they can be used to describe the possible shapes of wave functions of a particle in a box with definite energy, labelled by the quantum numbers $n = 1, 2, 3, \ldots$.

We conclude that the possible energy levels of a particle in a one-dimensional box of width a are given by

$$E_n = \frac{n^2 \pi^2 \hbar^2}{2ma^2}, \quad \text{with} \quad n = 1, 2, 3, \ldots, \tag{4.38}$$

and that a particle with energy E_n has a wave function of the form

$$\Psi_n(x, t) = N \sin k_n x \; e^{-iE_n t/\hbar}. \tag{4.39}$$

We note the following:

- As shown in Fig. 4.2, the separation between the energy levels increases as the quantum number n increases. However, this separation as a fraction of the energy decreases; indeed

$$\frac{E_{n+1} - E_n}{E_n} \to \frac{2}{n} \quad \text{as} \quad n \to \infty.$$

This means that the discrete nature of the energy levels becomes less important when the energy is high.

- The lowest possible energy, in contrast with classical physics, is not zero, but

$$E_1 = \frac{\hbar^2 \pi^2}{2ma^2}.$$

We can understand this *zero-point energy* by using the Heisenberg uncertainty principle, Eq. (1.17). If a particle is confined to a region of size a, it has an uncertain position $\Delta x \leq a$ and, hence, an uncertain momentum Δp which is at least of the order of $\hbar/2a$. Because the average magnitude of the momentum is always greater than Δp, the average kinetic energy of the particle is always greater than $(\Delta p)^2/2m$ which in turn is greater than $\hbar^2/8ma^2$.

- The spatial shape of the wave function of a particle in a box with energy E_n is identical to the spatial shape of the normal mode of a string with angular frequency ω_n. As illustrated in Fig. 4.1, the number of nodes increases as the value of n increases.

• The wave function of a particle in a box, unlike the displacement of a string, is not observable, but it can be used to construct properties of the particle that are observable. The first step is to normalize the wave function so that

$$\int_0^a |\Psi(x, t)|^2 \, dx = 1.$$

As shown in Section 3.4, this condition is satisfied if $N = \sqrt{2/a}$. One can then calculate probability densities for position and momentum as illustrated in Fig. 3.3.

A three-dimensional box

We shall now consider the more realistic problem of a particle confined in three dimensions. If the particle has definite energy E, its wave function has the form

$$\Psi(x, y, z, t) = \psi(x, y, z) \, e^{-iEt/\hbar}, \tag{4.40}$$

where $\psi(x, y, z)$ satisfies the energy eigenvalue equation

$$\left[-\frac{\hbar^2}{2m} \left(\frac{\partial^2}{\partial x^2} + \frac{\partial^2}{\partial y^2} + \frac{\partial^2}{\partial z^2} \right) + V(x, y, z) \right] \psi = E\psi. \tag{4.41}$$

We shall choose a potential energy function

$$V(x, y, z) = \begin{cases} 0 & \text{if } 0 < x < a, 0 < y < b, 0 < z < c \\ \infty & \text{elsewhere,} \end{cases} \tag{4.42}$$

which confines the particle to a box with sides a, b and c.

The possible energy eigenfunctions and eigenvalues of the particle may be found by seeking solutions of Eq. (4.41) inside the box which are equal to zero on all six faces of the box. For example, the function

$$\psi(x, y, z) = N \, \sin\left(\frac{\pi x}{a}\right) \sin\left(\frac{\pi y}{b}\right) \sin\left(\frac{\pi z}{c}\right)$$

is zero on each of the faces defined by

$$x = 0, \quad x = a, \quad y = 0, \quad y = b, \quad z = 0 \quad \text{and} \quad z = c,$$

and it satisfies Eq. (4.41) inside the box where $V(x, y, z) = 0$, if

$$E = \frac{\hbar^2 \pi^2}{2m} \left[\frac{1}{a^2} + \frac{1}{b^2} + \frac{1}{c^2} \right].$$

In general, there are an infinite set of eigenfunctions and eigenvalues labelled by three quantum numbers $n_x = 1, 2, 3, \ldots, n_y = 1, 2, 3, \ldots,$ and $n_z = 1, 2, 3, \ldots.$ The eigenfunctions have the form

$$\psi_{n_x, n_y, n_z}(x, y, z) = N \sin\left(\frac{n_x \pi x}{a}\right) \sin\left(\frac{n_y \pi y}{b}\right) \sin\left(\frac{n_z \pi z}{c}\right), \qquad (4.43)$$

and the energy eigenvalues are given by

$$E_{n_x, n_y, n_z} = \frac{\hbar^2 \pi^2}{2m} \left[\frac{n_x^2}{a^2} + \frac{n_y^2}{b^2} + \frac{n_z^2}{c^2} \right]. \qquad (4.44)$$

Equation (4.44) shows how the quantized energy levels of a particle in a box depend upon the dimensions of the box, a, b and c. Most importantly, it shows that some energy levels may coincide when the box has particular dimensions. We illustrate this in Fig. 4.3 which shows that, for a particle in a cubical box with $a = b = c$, energy levels like $E_{1,1,2}$, $E_{2,1,1}$, and $E_{1,2,1}$ coincide. When there are several states, or wave functions, with the same energy, the energy levels are said to be *degenerate*. Degenerate energy levels are very important in atomic, nuclear and particle physics. They arise because the interactions which confine

$$E_{2,2,2} = 16 \frac{\hbar^2 \pi^2}{2ma^2}$$

$$E_{1,1,3} = E_{3,1,1} = E_{1,3,1} = 11 \frac{\hbar^2 \pi^2}{2ma^2}$$

$$E_{1,2,2} = E_{2,2,1} = E_{2,1,2} = 9 \frac{\hbar^2 \pi^2}{2ma^2}$$

$$E_{1,1,2} = E_{2,1,1} = E_{1,2,1} = 6 \frac{\hbar^2 \pi^2}{2ma^2}$$

$$E_{1,1,1} = 3 \frac{\hbar^2 \pi^2}{2ma^2}$$

Fig. 4.3 Low-lying energy levels of a particle of mass m confined in a cubical box with sides of length a. Note the degeneracy of the first, second and third excited states.

electrons in atoms, nucleons in nuclei and quarks in hadrons, have specific symmetry properties. Indeed, the observed degeneracy of energy levels can be used to deduce these symmetry properties.

4.5 STATES OF UNCERTAIN ENERGY

In Section 4.3 we showed that a quantum state with a sharply defined energy E_n is represented by the wave function

$$\Psi_n(\mathbf{r}, t) = \psi_n(\mathbf{r})\, e^{-iE_n t/\hbar}, \tag{4.45}$$

where $\psi_n(\mathbf{r})$ is the energy eigenfunction belonging to the eigenvalue E_n. We shall now show that a state with uncertain energy is represented by a wave function of the form

$$\Psi(\mathbf{r}, t) = \sum_{n=1, 2, \ldots} c_n \psi_n(\mathbf{r})\, e^{-iE_n t/\hbar}. \tag{4.46}$$

To do so we need to understand two concepts. The mathematical concept of *a complete set of basis functions* and the physical concept of *an energy probability amplitude*.

Basis functions

Because the Schrödinger equation is a homogeneous linear partial differential equation, a linear superposition of wave functions with definite energies is also a solution. For example, the general wave function for a particle in a one-dimensional box is

$$\Psi(x, t) = \sum_{n=1, 2, 3, \ldots} c_n \Psi_n(x, t) \tag{4.47}$$

where, in accord with Eq. (4.39),

$$\Psi_n(x, t) = N \sin k_n x\; e^{-iE_n t/\hbar}$$

and c_n are arbitrary complex constants. This equation is very similar to the equation for the general motion of a vibrating string, Eq. (4.16). In one case, we have a general wave function expressed as a linear superposition of wave functions with definite energies and, in the other case, we have a general vibration expressed as a linear superposition of normal modes. In both these cases, we have a linear superposition of sine functions, or a Fourier sine series.

In more complicated problems, *a generalized Fourier series* must be used. Such a series, like an ordinary Fourier series, is based on the idea that the eigenfunctions of a Hamiltonian form a complete orthonormal set of basis functions. To illustrate this idea we shall consider a Hamiltonian which gives rise to an infinite set of energy eigenfunctions and eigenvalues denoted by $\psi_n(\mathbf{r})$ and E_n with $n = 1, 2, \ldots$. It can be shown that these eigenfunctions form a complete orthonormal set of basis functions. This means three things:

- Each eigenfunction $\Psi_n(\mathbf{r})$ may be normalized so that

$$\int |\psi_n(\mathbf{r})|^2 \, \mathrm{d}^3\mathbf{r} = 1. \tag{4.48}$$

- The eigenfunctions, ψ_m and ψ_n, belonging to different discrete eigenvalues E_m and E_n, are orthogonal. As shown in problem 2 at the end of this chapter, they satisfy the condition

$$\int \psi_m^*(\mathbf{r})\psi_n(\mathbf{r}) \, \mathrm{d}^3\mathbf{r} = 0 \quad \text{if} \quad E_m \neq E_n. \tag{4.49}$$

The eigenfunctions belonging to a degenerate eigenvalue, $E_m = E_n$, are not uniquely determined. One can utilize this latitude to make them orthogonal. The orthogonality relation (4.49) then holds generally.

- The eigenfunctions form a complete set because it is always possible to express any wave function as a linear superposition of eigenfunctions. This expression is a generalised Fourier series of the form

$$\Psi(\mathbf{r}, t) = \sum_{n=1, 2, 3\ldots} c_n \psi_n(\mathbf{r}) \, \mathrm{e}^{-iE_n t/\hbar}, \tag{4.50}$$

where the coefficients c_n are complex constants.

The coefficients c_n of the generalized Fourier series (4.50) can be found if we know the initial spatial shape of the wave function. For example, to find c_3 we consider

$$\Psi(\mathbf{r}, 0) = c_1\psi_1(\mathbf{r}) + c_2\psi_2(\mathbf{r}) + c_3\psi_3(\mathbf{r}) + \ldots,$$

and multiply both sides by $\psi_3^*(\mathbf{r})$. If we integrate over $\mathbf{r}$ and use the normalization and orthogonality integrals, Eqs. (4.48) and (4.49), we obtain

$$\int \psi_3^*(\mathbf{r})\Psi(\mathbf{r}, 0) \, \mathrm{d}^3\mathbf{r} = c_3.$$

Clearly, the general expression for the coefficients c_n is

$$c_n = \int \psi_n^*(\mathbf{r})\Psi(\mathbf{r}, 0)\, d^3\mathbf{r}. \tag{4.51}$$

Armed with these coefficients, we can use Eq. (4.50) to keep track of the subsequent evolution of the wave function.[1]

Energy probability amplitudes

We shall now consider the physical interpretation of a wave function given by the linear superposition of energy eigenfunctions,

$$\Psi(\mathbf{r}, t) = \sum_{n=1, 2, \ldots} c_n \psi_n(\mathbf{r})\, e^{-iE_n t/\hbar}. \tag{4.52}$$

Our first step is to find the condition for the wave function (4.52) to be normalized. The normalization integral $\int \Psi^*\Psi\, d^3\mathbf{r}$ has the form

$$\int [c_1^*\psi_1^*\, e^{+iE_1 t/\hbar} + c_2^*\psi_2^*\, e^{+iE_2 t/\hbar} + \ldots]\,[c_1\psi_1\, e^{-iE_1 t/\hbar} + c_2\psi_2\, e^{-iE_2 t/\hbar} + \ldots]\, d^3\mathbf{r}.$$

Using the normalization and othogonality relations for the eigefunctions $\psi_n(\mathbf{r})$, we find that terms like

$$c_2^*c_1\, e^{i(E_2 - E_1)t/\hbar} \int \psi_2^*\psi_1\, d^3\mathbf{r}$$

yield zero and terms like

$$c_1^*c_1 \int \psi_1^*\psi_1\, d^3\mathbf{r}$$

yield $|c_1|^2$. Hence, we obtain

$$\int \Psi^*\Psi\, d^3\mathbf{r} = \sum_{n=1, 2, \ldots} |c_n|^2,$$

and conclude that the wave function (4.52) is normalized if the coefficients c_n satisfy the condition

$$\sum_{n=1, 2, \ldots} |c_n|^2 = 1. \tag{4.53}$$

---

[1] These equations have to be modified when the potential gives rise to a continuum of energy eigenvalues. In this case, the normalization and orthogonality of the eigenfunctions involves a Dirac delta function and the general solution involves an integral over the continuous energy variable which labels the eigenfunctions. These modifications will be considered in Chapter 7.

We shall now calculate the expectation values for the energy and for the square of the energy of a particle with wave function (4.52). These, according to Eqs. (4.26) and (4.27), are given by

$$\langle E \rangle = \int \Psi^* \hat{H} \, \Psi \, d^3\mathbf{r} \quad \text{and} \quad \langle E^2 \rangle = \int \Psi^* \hat{H}^2 \, \Psi \, d^3\mathbf{r}.$$

To evaluate these integrals when the wave function is a superposition of eigenfunctions ψ_n, we use

$$\hat{H}\psi_n(\mathbf{r}) = E_n\psi_n(\mathbf{r}) \quad \text{and} \quad \hat{H}^2\psi_n(\mathbf{r}) = E_n^2\psi_n(\mathbf{r}),$$

and the orthogonality and normalization relations, Eqs. (4.48) and (4.49). We obtain

$$\langle E \rangle = \sum_{n=1,2,\dots} |c_n|^2 E_n \quad \text{and} \quad \langle E^2 \rangle = \sum_{n=1,2,\dots} |c_n|^2 E_n^2. \tag{4.54}$$

We can now assign a physical meaning to the complex coefficients c_n. If we recall the general statements made about probability distributions in Section 3.1, we see that

$$p_n = |c_n|^2 \quad \text{with} \quad n = 1,2,3\dots. \tag{4.55}$$

is a probability distribution. Eq. (4.53) shows that it is a normalized distribution and Eq. (4.54) shows that $p_n = |c_n|^2$ is the probability that the energy is equal to E_n. Because of this, the coefficients c_n are called *energy probability amplitudes*.

In this section we have illustrated how the possible energies of a quantum state are described using the principle of linear superposition. According to this principle all quantum states are linear superpositions of other quantum states. Here we have shown that the state represented by the linear superposition (4.52) describes a particle with uncertain energy and that the possible outcomes of a measurement are $E_1, E_2, E_3, \dots$ with probabilities $|c_1|^2, |c_2|^2, |c_3|^2, \dots$.

4.6   TIME DEPENDENCE

We shall end this chapter by exploring the fundamental connection between the energy properties and the time dependence of the observable properties of a quantum state.

We first consider a particle with the wave function

$$\Psi_n(\mathbf{r}, t) = \psi_n(\mathbf{r}) \, e^{-iE_n t/\hbar} \tag{4.56}$$

which represents a quantum state with sharply defined energy E_n. Even though the time dependence of this wave function is described by a complex exponential which oscillates with angular frequency $E_n/\hbar$, the observable properties of the particle do not change with time. We can illustrate this by showing that the position probability density is independent of time:

$$\Psi_n^*(\mathbf{r}, t)\Psi_n(\mathbf{r}, t) = \psi_n^*(\mathbf{r})\,e^{+iE_n t/\hbar}\psi_n(\mathbf{r})\,e^{-iE_n t/\hbar} = \psi_n^*(\mathbf{r})\psi_n(\mathbf{r}).$$

In fact, the wave function $\Psi_n(\mathbf{r}, t)$ describes a quantum state with no observable time dependence; the probabilities and the expectation value for any observable never change. Such a state is called a *stationary state*.

In contrast, a quantum state with uncertain energy has observable properties which change with time. We shall illustrate this by considering a particle with wave function

$$\Psi(\mathbf{r}, t) = \sqrt{\tfrac{1}{2}}\psi_1(\mathbf{r})\,e^{-iE_1 t/\hbar} + \sqrt{\tfrac{1}{2}}\psi_2(\mathbf{r})\,e^{-iE_2 t/\hbar}. \tag{4.57}$$

In this case, there are two possible outcomes when the energy is measured: E_1 with probability $\tfrac{1}{2}$ and E_2 with probability $\tfrac{1}{2}$. The energy expectation values are

$$\langle E \rangle = \tfrac{1}{2}E_1 + \tfrac{1}{2}E_2 \quad \text{and} \quad \langle E^2 \rangle = \tfrac{1}{2}E_1^2 + \tfrac{1}{2}E_2^2,$$

and the uncertainty in energy is

$$\Delta E = \sqrt{\langle E^2 \rangle - \langle E \rangle^2} = \tfrac{1}{2}|E_1 - E_2|.$$

The position probability density for this state of uncertain energy is time-dependent: it is given by

$$\Psi^*\Psi = \tfrac{1}{2}[|\psi_1|^2 + |\psi_2|^2 + \psi_1^*\psi_2\,e^{+i(E_1-E_2)t/\hbar} + \psi_1\psi_2^*\,e^{-i(E_1-E_2)t/\hbar}],$$

and oscillates with angular frequency $|E_1 - E_2|/\hbar$, i.e. with period $2\pi\hbar/|E_1 - E_2|$. Thus, Eq. (4.57) represents a quantum state with uncertain energy ΔE which has an observable property that oscillates with period $\pi\hbar/\Delta E$.

Quantum states of uncertain energy are called *non-stationary states* because they have some observable properties which change with time. In general, these properties change more rapidly when the energy is more uncertain. If δt is the time scale for significant change and ΔE is the energy uncertainty, then

$$\delta t\,\Delta E \approx \hbar. \tag{4.58}$$

This general relation is often referred to as the Heisenberg uncertainty relation for time and energy, but a better name is *the time and uncertainty in energy relation*. Time, unlike position, momentum or energy, is not an observable in quantum mechanics. It is a parameter which is used to label a changing system. The symbol δt in Eq. (4.58) is not an uncertainty in the outcome of a measurement, but the time scale for change in observable properties of a state. If this time scale is short, we have a non-stationary state with a large uncertainty in its energy. If this time scale is infinite, we have a stationary state with certain energy.

In practice, the distinction between stationary and non-stationary states is subtle. Clearly, the ground state of an atom is a state of definite energy and hence a stationary state with no observable time dependence; accordingly, the electrons in an undisturbed atom, even though they have kinetic energy, have no time-dependent properties. At first sight, an atom in an excited state is also in a state of definite energy; indeed, its wave function is an energy eigenfunction of the Hamiltonian which describes the interactions between the particles inside the atom. As such, the atom should be in a stationary state, a state with timeless properties. But an atom in an excited state changes; it emits electromagnetic radiation and de-excites. It is, at best, in an almost stationary state and, according to the uncertainty relation (4.58), its energy must have a small uncertainty.

In fact, an excited state of an atom has uncertain energy because the true Hamiltonian describes not only the interactions between the particles inside the atom but also an interaction between these particles and fluctuating electromagnetic fields that are always present, even in empty space. These interactions give rise to an energy uncertainty ΔE which, in accord with Eq. (4.58), is given by

$$\Delta E = \frac{\hbar}{\tau},\tag{4.59}$$

where τ is the mean lifetime for the excited state to decay. As a consequence, the wavelength of radiation emitted by a decaying atom is uncertain and the observed spectral line has a *natural line width*. But, in most situations, natural line widths are smaller than the widths that arise because atoms move and collide.

Finally, the reader may be concerned that time and position are treated so differently in quantum mechanics; time is a label whereas position is an observable which can be measured. To treat them differently is against the spirit of special relativity which stresses the unity of time and space. Indeed, this shortcoming in quantum mechanics has to be rectified before the theory can be effective in describing relativistic phenomena. In relativistic quantum physics neither position nor time are observables; both are labels assigned to quantum field operators.

PROBLEMS 4

1. In practice the potential energy function for a particle is a real function. In this problem you are asked to show that this implies that the energy eigenvalues of a particle are real.

   The energy eigenfunction $\psi_n(x)$ and its complex conjugate $\psi_n^*(x)$ satisfy the equations

   $$\left[-\frac{\hbar^2}{2m}\frac{d^2}{dx^2} + V(x)\right]\psi_n = E_n\psi_n,$$

   and

   $$\left[-\frac{\hbar^2}{2m}\frac{d^2}{dx^2} + V(x)\right]\psi_n^* = E_n^*\psi_n^*.$$

   Multiply the first equation by ψ_n^* and the second by ψ_n, subtract and show that

   $$-\frac{\hbar^2}{2m}\frac{d}{dx}\left[\psi_n^*\frac{d\psi_n}{dx} - \psi_n\frac{d\psi_n^*}{dx}\right] = (E_n - E_n^*)\psi_n^*\psi_n.$$

   By integrating over x and by assuming that $\psi_n(x)$ is zero at $x = \pm\infty$, show that $E_n = E_n^*$.

2. In this problem you are asked to show that the eigenfunctions, ψ_m and ψ_n, belonging to different discrete eigenvalues E_m and E_n, are orthogonal.

   For a real potential, the energy eigenfunction $\psi_n(x)$ and the complex conjugate of $\psi_m(x)$ satisfy the equations

   $$\left[-\frac{\hbar^2}{2m}\frac{d^2}{dx^2} + V(x)\right]\psi_n = E_n\psi_n,$$

   and

   $$\left[-\frac{\hbar^2}{2m}\frac{d^2}{dx^2} + V(x)\right]\psi_m^* = E_m\psi_m^*.$$

   Multiply the first equation by ψ_m^* and the second by ψ_n, subtract and show that

$$-\frac{\hbar^2}{2m}\frac{d}{dx}\left[\psi_m^*\frac{d\psi_n}{dx} - \psi_n\frac{d\psi_m^*}{dx}\right] = (E_n - E_m)\psi_m^*\psi_n.$$

By integrating over x and by assuming that $\psi_n(x)$ and $\psi_m(x)$ are zero at $x = \pm\infty$, show that

$$\int_{-\infty}^{+\infty} \psi_m^*(x)\psi_n(x)\,dx = 0 \quad \text{if} \quad E_m \neq E_n.$$

3. What is the energy difference between the lowest and first excited state of a particle of mass m in a one-dimensional, infinite square-well of width a? Evaluate this energy:

(a) for an electron in a well of atomic size, expressing your answer in eV,

(b) for a neutron in a well of nuclear size, expressing your answer in MeV.

4. According to Eq. (4.39), the wave functions for a particle with energy E_n confined by an infinite square-well potential are

$$\Psi_n(x, t) = N \sin k_n x \; e^{-iE_n t/\hbar}.$$

Show that these wave functions can be thought of as standing waves formed by a linear superposition of travelling waves trapped in the region $0 < x < a$.

5. Consider a particle of mass m in a two-dimensional box defined by the potential energy field:

$$V(x, y) = \begin{cases} 0 & \text{if } 0 < x < a \text{ and } 0 < y < b \\ \infty & \text{elsewhere.} \end{cases}$$

States of definite energy can be labelled by two quantum numbers, n_x and n_y, and they have wave functions of the form

$$\Psi_{n_x, n_y}(x, y, t) = \psi_{n_x, n_y}(x, y) \; e^{-iE_{n_x, n_y} t/\hbar}.$$

(a) Find the explicit form of the eigenfunctions $\psi_{n_x, n_y}(x, y)$ and eigenvalues E_{n_x, n_y}.

(b) Draw diagrams showing the first four energy levels in a box with $a = b$ and in a box with $a = 2b$. Indicate on the diagrams the degeneracy of each level, i.e. the number of independent eigenfunctions that have the same energy.

6. Two states of a particle with definite energy E_1 and E_2 are represented by the following normalized, orthogonal solutions of the Schrödinger equation:

$$\Psi_1(x, t) = \psi_1(x) e^{-iE_1 t/\hbar} \quad \text{and} \quad \Psi_2(x, t) = \psi_2(x) e^{-iE_2 t/\hbar}.$$

(a) Write down a linear superposition of Ψ_1 and Ψ_2 which represents the state for which the expectation value of the energy is $\frac{1}{4} E_1 + \frac{3}{4} E_2$.

(b) Find the uncertainty in energy for the state written down.

(c) Show, for the state written down, that the probability density oscillates with time. Find the relation between the period of these oscillations and the uncertainty in the energy.

7. This problem illustrates the general idea of a complete set of orthonormal basis functions on the interval $0 < x < a$ by considering the eigenfunctions of a particle in a one-dimensional infinite square well of width a. These are given by Eq. (4.37), i.e. by

$$\psi_n(x) = N \sin k_n x \quad \text{for} \quad 0 < x < a,$$

where

$$k_n = \frac{n\pi}{a} \quad \text{and} \quad n = 1, 2, 3 \dots.$$

(a) Show that the normalization condition

$$\int_0^a |\psi_n(x)|^2 \, dx = 1$$

is satisfied if $N = \sqrt{2/a}$.

(b) Show that the orthogonality condition

$$\int_0^a \psi_m^* \psi_n \, dx = 0 \quad \text{if} \quad m \neq n$$

is satisfied.

(c) Consider the Fourier sine series for the function $f(x)$ on the interval $0 < x < a$

$$f(x) = \sum_{n=1, 2, 3\dots} c_n \psi_n(x).$$

Show that the coefficients of this series are given by

$$c_n = \frac{2}{a} \int_0^a \sin k_n x \, f(x) \, dx.$$

8. Consider a particle of mass m in the ground state of an infinite square-well potential of width $a/2$. Its normalized wave function at time $t = 0$ is

$$\Psi(x,0) = \begin{cases} \dfrac{2}{\sqrt{a}} \sin \dfrac{2\pi x}{a} & \text{if } 0 < x < a/2 \\ 0 & \text{elsewhere.} \end{cases}$$

At this time the well suddenly changes to an infinite square-well potential of width a without affecting the wave function.

By writing $\Psi(x,t)$ as a linear superposition of the energy eigenfunctions of the new potential, find the probability that a subsequent measurement of the energy will yield the result

$$E_1 = \frac{\hbar^2 \pi^2}{2ma^2}.$$

(Hint: A linear superposition of square-well eigenfunctions is a Fourier sine series, and the coefficients of the series are given by simple integrals.)

9. The general wave function of a particle of mass m in a one-dimensional infinite square well with width a at time t is

$$\Psi(x,t) = \sum_{n=1}^{\infty} c_n \psi_n(x) e^{-iE_n t/\hbar},$$

where $\psi_n(x)$ is an eigenfunction with energy $E_n = n^2 \pi^2 \hbar^2 / 2ma^2$. Show that the wave function returns to its original form in a time $T = 4ma^2/\pi\hbar$.

10. Consider the electromagnetic radiation of wavelength λ which is emitted when an atom makes a transition from a state with energy E_2 to a ground state with energy E_1. Assume that the mean lifetime of the state with energy E_2 is τ. Show that the uncertainty in the wavelength of the emitted radiation, i.e. the natural line width of the spectral line, is given by

$$\Delta\lambda = \frac{\lambda^2}{2\pi c\tau}.$$

11. In particle physics the Z boson is an unstable gauge boson which plays a key role in mediating the Weak Nuclear Interaction. The fundamental uncertainty in the mass energy of the Z boson is $\Delta E = 2.5\,\text{GeV}$. Evaluate the mean decay lifetime of the Z boson.

5

Square wells and barriers

Insight into how quantum particles can be bound or scattered by potential energy fields can be obtained by considering models based on square wells and square barriers. In these models, the Schrödinger equation may be solved easily using elementary mathematics, the possible energies of a particle may be found and the properties of the wave functions are self-evident.

We begin by considering the quantum states of a particle in a one-dimensional square-well potential. We shall show that there are unbound states with a continuous range of energies and that there are, when the well is deep enough, bound states with discrete energies.

We shall then consider a particle incident on a square potential barrier. We shall see that this is an uncertain encounter with two possible outcomes: reflection and transmission. Most importantly, we shall show that transmission is possible even when the particle has insufficient energy to surmount the barrier. In other words, we shall illustrate how quantum particles can tunnel through potential barriers.

5.1 BOUND AND UNBOUND STATES

In order to explore the properties of bound and unbound quantum states in a simple context, we shall consider a particle of mass m in the one-dimensional potential energy field given by

$$V(x) = \begin{cases} \infty & \text{if } -\infty < x < 0 \\ -V_0 & \text{if } 0 < x < a \\ 0 & \text{if } a < x < \infty. \end{cases} \qquad (5.1)$$

As illustrated in Fig. 5.1, the potential energy changes abruptly at $x = 0$ and $x = a$. There is an attractive well of depth V_0, which may or may not trap the particle, and an infinite wall at $x = 0$ which repels the particle.

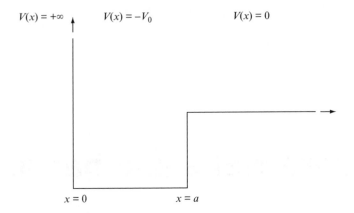

Fig. 5.1   The potential energy field given by Eq. (5.1) in which there are unbound states with a continuous range of energies and, if the well is deep enough, bound states with discrete energies.

The behaviour of a classical particle in this potential should be familiar. The energy of the particle, E, is given by the sum of its kinetic and potential energies,

$$E = \frac{p^2}{2m} + V(x).$$

When the energy is negative and somewhere between $E = -V_0$ and $E = 0$, the particle is bound or trapped in the well of depth V_0; it bounces back and forth between $x = 0$ and $x = a$ with kinetic energy $E + V_0$. But when the energy is positive, the particle is unbound. For example, it could approach the well from $x = +\infty$ with kinetic energy E, increase its kinetic energy to $E + V_0$ when it reaches $x = a$, hit the infinitely-high potential wall at $x = 0$ and then bounce back to $x = +\infty$.

The behaviour of a quantum particle in this potential is described by a wave function $\Psi(x, t)$ which is a solution of the Schrödinger equation

$$i\hbar \frac{\partial \Psi}{\partial t} = -\frac{\hbar^2}{2m} \frac{\partial^2 \Psi}{\partial x^2} + V(x)\Psi. \qquad (5.2)$$

When the particle has definite energy E, the wave function has the form

$$\Psi(x, t) = \psi(x)\,e^{-iEt/\hbar}, \qquad (5.3)$$

where $\psi(x)$ is an eigenfunction satisfying the energy eigenvalue equation

$$-\frac{\hbar^2}{2m} \frac{d^2 \psi}{dx^2} + V(x)\psi = E\psi. \qquad (5.4)$$

Once we have solved this eigenvalue equation and found all the possible energy eigenvalues and eigenfunctions, we can represent any quantum state of the particle in the potential as a linear superposition of energy eigenfunctions.

To solve the eigenvalue equation, we note that the potential $V(x)$ given by Eq. (5.1) takes on constant values in three regions of x: $(-\infty < x < 0)$, $(0 < x < a)$ and $(a < x < +\infty)$. We shall find solutions of Eq. (5.4) in these three regions and then join the solutions together at $x = 0$ and at $x = a$ to obtain physically acceptable eigenfunctions. Because the potential $V(x)$ changes abruptly at $x = 0$ and $x = a$, we can only require the eigenfunctions $\psi(x)$ to be as smooth as possible. In particular, we shall require $\psi(x)$ to be continuous at $x = 0$ and $x = a$ in order to avoid unacceptable abrupt changes in the position probability density. The differential equation (5.4) with potential (5.1) then implies that the first derivative of $\psi(x)$ is continuous at $x = a$ and discontinuous at $x = 0$.[1]

We are interested in two types of eigenfunctions. The eigenfunctions for bound states and the eigenfunctions for unbound states.

Bound states

If a bound state exists, it has a negative energy somewhere between $E = -V_0$ and $E = 0$. We shall set $E = -\epsilon$, where ϵ is the binding energy, and seek solutions of Eq. (5.4).

In the region $(-\infty < x < 0)$, the potential energy is infinite and the only finite solution of Eq. (5.4) is $\psi(x) = 0$, signifying that the particle is never found in the negative x region.

In the region $(0 < x < a)$, the potential energy is $V(x) = -V_0$ and Eq. (5.4) has the form

$$\frac{d^2\psi}{dx^2} = -k_0^2\psi, \quad \text{where} \quad E = \frac{\hbar^2 k_0^2}{2m} - V_0. \tag{5.5}$$

The general solution of this second-order differential equation has the form

$$\psi(x) = C\sin(k_0 x + \gamma),$$

where C and γ are arbitrary constants. To ensure continuity of $\psi(x)$ at $x = 0$, we shall set the constant γ to zero to give

$$\psi(x) = C\sin k_0 x. \tag{5.6}$$

---

[1] The infinite change in the potential at $x = 0$ forces a discontinuity in $d\psi/dx$ at $x = 0$. A more rigorous approach would be to consider a potential energy with a finite value V_1 in the region $x \leq 0$, require the continuity of $\psi(x)$ and $d\psi/dx$ at $x = 0$ and then take the limit $V_1 \to \infty$.

In the region ($a < x < +\infty$), the potential energy is zero and Eq. (5.4) has the form

$$\frac{d^2\psi}{dx^2} = \alpha^2\psi, \quad \text{where} \quad E = -\frac{\hbar^2\alpha^2}{2m}. \tag{5.7}$$

The general solution is

$$\psi(x) = A\,e^{-\alpha x} + A'\,e^{+\alpha x},$$

where A and A' are arbitrary constants. To ensure that the eigenfunction is finite at infinity, we set A' to zero to give a solution which falls off exponentially with x:

$$\psi(x) = A\,e^{-\alpha x}. \tag{5.8}$$

Our next task is to join the solution given by Eq. (5.6), which is valid in the region ($0 < x < a$), onto the solution given by Eq. (5.8), which is valid in the region ($a < x < +\infty$). As mentioned earlier, we shall require the eigenfunction and its first derivative to be continuous at $x = a$. Continuity of $\psi(x)$ gives

$$C \sin k_0 a = A\,e^{-\alpha a}, \tag{5.9}$$

and continuity of $d\psi/dx$ gives

$$k_0 C \cos k_0 a = -\alpha A\,e^{-\alpha a}. \tag{5.10}$$

If we divide Eq. (5.10) by Eq. (5.9), we obtain

$$k_0 \cot k_0 a = -\alpha. \tag{5.11}$$

Equation (5.11) sets the condition for a smooth join at $x = a$ of the functions $C \sin k_0 x$ and $A\,e^{-\alpha x}$. It is a non-trivial condition which is only satisfied when the parameters k_0 and α take on special values. And once we have found these special values, we will be able to find the binding energies of the bound states from $\epsilon = \hbar^2\alpha^2/2m$.

To find these binding energies, we note that α and k_0 are not independent parameters. They are defined by

$$E = -\frac{\hbar^2\alpha^2}{2m} \quad \text{and} \quad E = \frac{\hbar^2 k_0^2}{2m} - V_0,$$

which imply that

$$\alpha^2 + k_0^2 = w^2, \quad \text{where } w \text{ is given by} \quad V_0 = \frac{\hbar^2 w^2}{2m}. \tag{5.12}$$

Thus, we have two simultaneous equations for α and k_0, Eq. (5.11) and Eq. (5.12). These equations may be solved graphically by finding the points of intersection of the curves

$$\alpha = -k_0 \cot k_0 a \quad \text{and} \quad \alpha^2 + k_0^2 = w^2,$$

as illustrated in Fig. 5.2.

Inspection of Fig. 5.2 shows the number of points of intersection, and hence the number of bound states, increase as the well becomes deeper. In particular, there are no bound states for a shallow well with

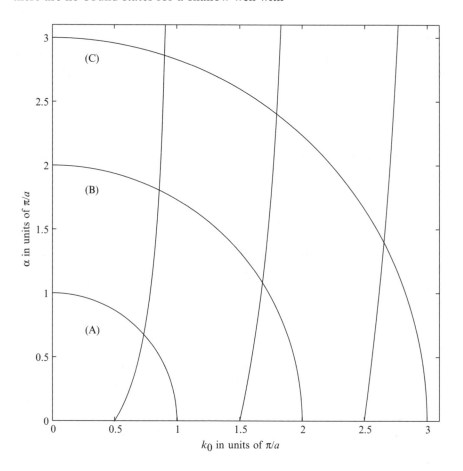

Fig. 5.2 Graphical solution of the simultaneous equations $\alpha = -k_0 \cot k_0 a$ and $\alpha^2 + k_0^2 = w^2$. The units of k_0 and α are π/a. Three values for the well-depth parameter, $w = \pi/a$, $w = 2\pi/a$ and $w = 3\pi/a$, are labelled by (A), (B) and (C), respectively. For (A) there is one point of intersection and one bound state, for (B) there are two points of intersection and two bound states and for (C) there are three points of intersection and three bound states.

$$w < \frac{\pi}{2a}.$$

There is one bound state when

$$\frac{\pi}{2a} < w < \frac{3\pi}{2a},$$

and two bound states when

$$\frac{3\pi}{2a} < w < \frac{5\pi}{2a},$$

and so on.

To illustrate the nature of the bound states, we shall consider a potential with well-depth parameter $w = 2\pi/a$ which corresponds to a well with depth

$$V_0 = \frac{2\hbar^2\pi^2}{ma^2}.$$

In this case two bound states exist, a ground state and a first excited state with binding energies

$$\epsilon_1 = 3.26 \, \frac{\hbar^2\pi^2}{2ma^2} \quad \text{and} \quad \epsilon_2 = 1.17 \, \frac{\hbar^2\pi^2}{2ma^2}.$$

The corresponding eigenfunctions are shown in Fig. 5.3. These eigenfunctions show that a bound quantum particle can be found outside the classical region of confinement ($0 < x < a$). Specifically, for $x > 0$, bound-state eigenfunctions have non-zero values given by

$$\psi(x) = A\,\mathrm{e}^{-\alpha x}.$$

Hence the position probability densities for a bound particle fall off exponentially as x penetrates into the classically forbidden region. Because the parameter α is related to the binding energy ϵ via $\epsilon = \hbar^2\alpha^2/2m$, the degree of quantum penetration into the classically forbidden region is more pronounced when the binding energy is low. The phenomenon of quantum penetration will be considered further in Section 5.2.

Unbound states

We shall now consider a particle with positive energy E that approaches the well, shown in Fig. 5.1, from the right and is then reflected at $x = 0$. It is useful to write the energy of the particle as

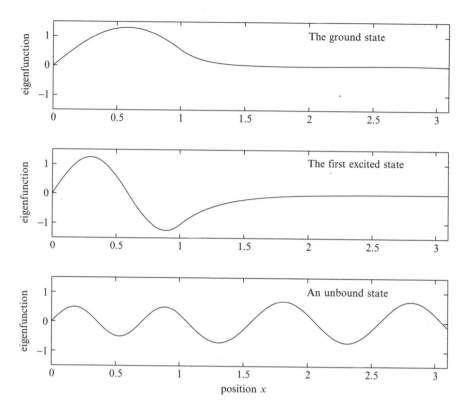

Fig. 5.3   The energy eigenfunctions for a particle in a square-well potential with width $a = 1$ and depth $V_0 = 2\hbar^2\pi^2/ma^2$. The eigenfunctions for the ground state, the first excited state, and an unbound state with energy $E = 2\hbar^2\pi^2/ma^2$ are shown. The normalization of the eigenfunction of the unbound state is arbitrary.

$$E = \frac{\hbar^2 k_0^2}{2m} - V_0 = \frac{\hbar^2 k^2}{2m} \tag{5.13}$$

and note that, if the particle were governed by classical physics, its momentum would be $\hbar k_0$ inside the well and $\hbar k$ outside the well. But the particle is really governed by quantum mechanics. It has a wave function of the form

$$\Psi(x,\ t) = \psi(x)\,e^{-iEt/\hbar}$$

where the eigenfunction $\psi(x)$ satisfies the eigenvalue equation (5.4) for all values of x. We shall find these eigenfunctions by following the procedure we used to find the bound-state eigenfunctions: we shall find solutions of Eq. (5.4) in the three regions of x and then join these solutions at $x = 0$ and $x = a$.

First, the eigenfunction $\psi(x)$ is zero in the region $(-\infty < x < 0)$ because the potential energy is infinite in this region.

Second, in the region $(0 < x < a)$ where the potential energy is $-V_0$, the eigenvalue equation for the unbound state has the same form as the eigenvalue equation for a bound state, Eq. (5.5), and the solution is given by Eq. (5.6), i.e. by

$$\psi(x) = C \sin k_0 x. \tag{5.14}$$

Third, in the region $(a < x < +\infty)$ where the potential energy is zero, the eigenvalue equation has the form

$$\frac{d^2\psi}{dx^2} = -k^2\psi, \tag{5.15}$$

and the solution is

$$\psi(x) = D \sin(kx + \delta). \tag{5.16}$$

The constant δ is called *the phase shift*.[2]

Our next task is to join the solution given by Eq. (5.14) onto the solution given by Eq. (5.16). As mentioned earlier, we shall require the eigenfunction and its first derivative to be continuous at $x = a$. Continuity of $\psi(x)$ gives

$$C \sin k_0 a = D \sin(ka + \delta) \tag{5.17}$$

and continuity of $d\psi/dx$ gives

$$k_0 C \cos k_0 a = kD \cos(ka + \delta). \tag{5.18}$$

If we divide Eq. (5.18) by Eq. (5.17), we obtain

$$k_0 \cot k_0 a = k \cot(ka + \delta). \tag{5.19}$$

This equation sets the condition for a smooth join of the functions $C \sin(k_0 x)$ and $D \sin(kx + \delta)$ at $x = a$. When we considered the analogous equation for the eigenfunctions of the bound states, Eq. (5.11), we found that the join was only smooth when the energy takes on special discrete values. This is not the case for unbound states. Here we can choose any value of the energy E, find k

---

[2] We recall that the bound-state eigenfunction, Eq. (5.8), decreases exponentially at large x signifying that a bound particle does not escape to infinity. In contrast, an unbound particle can escape to infinity and its eigenfunction has a finite value at large distances given by Eq. (5.16). Readers may worry about whether an eigenfunction for an unbound particle can be normalized. This can be done by the mathematical trick of placing the entire system, the potential and the particle, in a big box; to some extent, this box represents the laboratory containing the system. In practice, the normalization of unbound quantum states is seldom considered. These states are usually characterized by a particle flux or by a probability current density.

and k_0 from Eq. (5.13) and use Eq. (5.19) to find the phase shift δ. Thus, for any positive energy E, we can always find a smooth eigenfunction of the form:

$$\psi(x) = \begin{cases} 0 & \text{if } -\infty < x < 0 \\ C\sin(k_0 x) & \text{if } 0 < x < a \\ D\sin(kx + \delta) & \text{if } a < x < \infty. \end{cases} \tag{5.20}$$

We note that there are undulations with wave numbers k_0 and k in regions where a classical particle would have momenta $\hbar k_0$ and $\hbar k$. These undulations are illustrated in Fig. 5.3 which shows the eigenfunction for an unbound particle with energy $E = 2\hbar^2\pi^2/ma^2$ in a potential well of depth $V_0 = 2\hbar^2\pi^2/ma^2$.

The eigenfunction Eq. (5.20) may be rewritten to reveal a wave being reflected by the potential well. To identify incident and reflected components, we rewrite the eigenfunction in terms of complex exponentials. If we use

$$\sin\theta = \frac{e^{+i\theta} - e^{-i\theta}}{2i}$$

and if we introduce two new constants

$$A_0 = -\frac{C}{2i} \quad \text{and} \quad A = -\frac{D\,e^{-2i\delta}}{2i},$$

we obtain

$$\psi(x) = \begin{cases} A_0\,e^{-ik_0 x} - A_0\,e^{+ik_0 x} & \text{if } 0 < x < a \\ A\,e^{-ikx} - A\,e^{2i\delta}\,e^{+ikx} & \text{if } a < x < \infty. \end{cases} \tag{5.21}$$

We can now identify a wave travelling back and forth with wave number k_0 inside the well, and, outside the well, we can identify incoming and outgoing waves with wave number k but with a phase difference of 2δ. These waves represent a particle with momentum $\hbar k$ being reflected by the potential well between $x = 0$ and $x = a$ and by the barrier at $x = 0$. Because the incoming and reflected waves have the same intensity $|A|^2$, the reflection is complete. The meaning of the phase shift will become evident in a moment.

Even though we have used words like *travelling back and forth*, *incoming* and *outgoing*, the eigenfunction given by Eq. (5.21) describes a quantum state which exhibits no observable time dependence because it is a state with definite energy E. To describe the *dynamics* of particle reflection properly, we need a non-stationary quantum state of uncertain energy. Such a state can be represented in the region $(a < x < \infty)$ by an incoming wave packet of the form

$$\Psi_i(x,\ t) = \int_0^\infty c(E')e^{-i(E't+p'x)/\hbar}\ dE', \quad \text{where} \quad E' = \frac{p'^2}{2m},$$

and an outgoing wave packet of the form

$$\Psi_f(x,\ t) = \int_0^\infty c(E')e^{-i(E't-p'x)/\hbar}\ e^{2i\delta(E')}\ dE'.$$

The function $c(E')$ is an energy probability amplitude and $|c(E')|^2 dE'$ is the probability that the particle has energy between E' and $E' + dE'$. If the function $|c(E')|^2$ is strongly peaked at $E' = E$, the wave packets represent an incoming and outgoing particle with energy E, momentum $p = \sqrt{2mE}$ and velocity $v = p/m$. In this case, one can show that

$$\Psi_i(x,\ t) \propto F(t + x/v) \tag{5.22}$$

and

$$\Psi_f(x,\ t) \propto F(t - x/v - 2\hbar d\delta/dE), \tag{5.23}$$

where the function F specifies the shape of the incoming and outgoing wave packets. These equations show that the energy dependence of the phase shift, $\delta(E)$, is related to the time delay that occurs during the reflection process; this delay is approximately given by

$$\tau \approx 2\hbar \frac{d\delta}{dE}. \tag{5.24}$$

This time delay is negative for a particle reflected by the potential shown in Fig. 5.1 because the particle speeds up as it enters the well at $x = a$.

For the benefit of more advanced readers, we shall give a brief derivation of the formula for the time delay. If $c(E')$ is strongly peaked at $E' = E$, we can set

$$p' = p + (E' - E)\frac{dp}{dE}$$

and

$$\delta(E') = \delta(E) + (E' - E)\frac{d\delta}{dE},$$

and substitute into Eq. (5.22) and Eq. (5.23). If we note that dp/dE is equal to $1/v$, where v is the classical velocity of the particle, we obtain

$$\Psi_i(x, \ t) = e^{-i(Et+px)/\hbar} \int_0^\infty c(E')\, e^{-i(E'-E)(t+x/v)/\hbar}\ \mathrm{d}E'$$

and

$$\Psi_f(x, \ t) = e^{-i(Et-px-2\hbar\delta)/\hbar} \int_0^\infty c(E')\, e^{-i(E'-E)(t-x/v-2\hbar\mathrm{d}\delta/\mathrm{d}E)/\hbar}\ \mathrm{d}E'.$$

We now focus on the x and t dependence that emerges when the integration over E' is carried out. It should be clear that, for a given function $c(E')$, integration will lead to wave packets of a specific shape, which we denote by a function F, with locations given by

$$\Psi_i(x, \ t) \propto F(t + x/v)$$

and

$$\Psi_f(x, \ t) \propto F(t - x/v - 2\hbar\mathrm{d}\delta/\mathrm{d}E).$$

Thus, we have an incoming wave packet with velocity v and, after a time delay of $2\hbar\mathrm{d}\delta/\mathrm{d}E$, an outgoing wave packet with velocity v. In this approximate calculation the shapes of the incoming and outgoing wave packets are the same, but in practice wave packets change in shape as they move.

General implications

This section has considered a particle in the potential energy field given in Fig. 5.1. We have chosen this simple potential because it permits the solution of the Schrödinger equation using elementary methods and because it leads to wave functions which are easy to visualize. Clearly, the detail of the results are only relevant to this particular potential, but the following general features are relevant in atomic, nuclear and particle physics:

- Wave functions undulate in classically allowed regions and fall off exponentially in classically forbidden regions.

- Potentials give rise to bound states with discrete energies when they are sufficiently attractive.

- When a particle is unbound it can have a continuous range of energies, and when it is scattered or reflected by a potential, its wave function experiences a phase shift which can be related to a time delay that occurs during the scattering process.

5.2   BARRIER PENETRATION

The ability to penetrate and tunnel through a classically forbidden region is one of the most important properties of a quantum particle. In the last section, we discovered that the wave function of a bound particle extends beyond the region of confinement of a bound classical particle. In this section we shall show how particles can tunnel through potential barriers. To keep the mathematics as simple as possible we shall consider a particle in a simple potential energy field of the form:

$$V(x) = \begin{cases} 0 & \text{if } -\infty < x < 0 \\ V_B & \text{if } 0 < x < a \\ 0 & \text{if } a < x < +\infty. \end{cases} \qquad (5.25)$$

As shown in Fig. 5.4, we have a square barrier of height V_B separating the regions $(-\infty < x < 0)$ and $(a < x < +\infty)$.

If a classical particle were to approach this barrier from the left, it would be reflected if its energy is below V_B and it would be transmitted if its energy is above V_B. We shall see that when a quantum particle encounters the barrier, the outcome is uncertain; it may be reflected or it may be transmitted. Most importantly, we shall show that the particle may be transmitted even when its energy is below V_B, and we shall calculate the probability for this to happen.

The behaviour of a particle of mass m in the potential $V(x)$ is described by a wave function $\Psi(x, t)$ which is a solution of the Schrödinger equation

$$i\hbar \frac{\partial \Psi}{\partial t} = -\frac{\hbar^2}{2m} \frac{\partial^2 \Psi}{\partial x^2} + V(x)\Psi.$$

To describe the dynamics of the uncertain encounter with the barrier, we seek a wave function $\Psi(x, t)$ which describes an incoming particle and the possibility of reflection and transmission. This wave function should give rise to an incoming pulse of probability representing a particle approaching the barrier

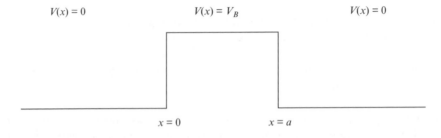

Fig. 5.4   The potential barrier given by Eq. (5.25) which will be used to illustrate quantum mechanical tunnelling.

before the encounter. After the encounter, there should be two pulses of probability, one representing the possibility of a reflected particle and the other the possibility of a transmitted particle. According to the standard interpretation of the wave function, reflection and transmission persist as possible options until the particle is detected. When this happens, the wave function $\Psi(x, t)$ collapses and one or other of the two options are realized with probabilities governed by the magnitudes of the reflected and transmitted pulses.

This dynamical description suggests that we need to solve a problem in time-dependent quantum mechanics. The quantum state that describes an incoming particle and the possibilities of a reflected and transmitted particle is a non-stationary state. Such a state, as discussed in Section 4.7, is a state of uncertain energy represented by a linear superposition of energy eigenfunctions. In this case, there is a continuum of possible energies and the wave function has the form

$$\Psi(x, t) = \int c(E')\psi_{E'}(x)\,e^{-iE't/\hbar}\,dE',$$

where $\psi_{E'}(x)$ is an eigenfunction with energy E' with the appropriate boundary conditions. The function $c(E')$ is an energy probability amplitude and $|c(E')|^2\,dE'$ is the probability that the particle has energy between E' and $E' + dE'$. If the function $|c(E')|^2$ is strongly peaked at $E' = E$, the wave function $\Psi(x, t)$ consists of a localized wave packet representing a particle of energy E encountering the barrier.

Even though time-dependent quantum mechanics is essential for a conceptual understanding of an uncertain encounter with a potential barrier, it is not needed to calculate the probabilities of reflection and transmission. We shall now show how these probabilities may be found simply by using time-independent quantum mechanics.

Stationary state analysis of reflection and transmission

Provided the uncertainty in the energy of the particle is small compared with the variations in the potential energy $V(x)$, we can calculate the probabilities of reflection and transmission by considering a stationary state with definite energy. Such a state is represented by the wave function

$$\Psi(x, t) = \psi_E(x)\,e^{-iEt/\hbar}, \tag{5.26}$$

where $\psi_E(x)$ is an eigenfunction with energy E satisfying the eigenvalue equation

$$-\frac{\hbar^2}{2m}\frac{d^2\psi_E}{dx^2} + V(x)\psi_E = E\psi_E. \qquad (5.27)$$

Because of the simple nature of the potential $V(x)$ shown in Fig. 5.4, it is easy to find an eigenfunction which describes incoming, reflected and transmitted waves. The procedure is to find the solutions of Eq. (5.27) in different regions of x and then smoothly join the solutions together.[3]

On the left of the barrier, the potential energy $V(x)$ is zero and the eigenfunction $\psi_E(x)$ satisfies the differential equation

$$\frac{d^2\psi_E}{dx^2} = -k^2\psi_E, \quad \text{where} \quad E = \frac{\hbar^2 k^2}{2m}. \qquad (5.28)$$

The solution representing an incident wave of intensity $|A_I|^2$ and a reflected wave of intensity $|A_R|^2$ is

$$\psi_E(x) = A_I\,e^{+ikx} + A_R\,e^{-ikx}. \qquad (5.29)$$

The form of the eigenfunction inside the barrier depends on whether the energy of the particle is above or below the barrier. When $E > V_B$, the region $(0 < x < a)$ is a classically allowed region and the eigenfunction is governed by

$$\frac{d^2\psi_E}{dx^2} = -k_B^2\psi_E, \quad \text{where} \quad E = \frac{\hbar^2 k_B^2}{2m} + V_B. \qquad (5.30)$$

The general solution involves two arbitrary constants and it undulates with wave number k_B as follows

$$\psi_E(x) = A\,e^{+ik_Bx} + A'\,e^{-ik_Bx}. \qquad (5.31)$$

When $E < V_B$, the region $(0 < x < a)$ is a classically forbidden region. Here, the eigenfunction is governed by

$$\frac{d^2\psi_E}{dx^2} = \beta^2\psi_E, \quad \text{where} \quad E = -\frac{\hbar^2\beta^2}{2m} + V_B, \qquad (5.32)$$

and the general solution is

$$\psi_E(x) = B\,e^{-\beta x} + B'\,e^{+\beta x}, \qquad (5.33)$$

---

[3] A similar procedure is used to describe the reflection and transmission of classical waves. Indeed, the mathematics describing an electromagnetic wave incident on a thin sheet of metal is almost identical.

where B and B' are arbitrary constants.

The potential energy is again equal to zero on the right of the barrier. Here, the eigenfunction satisfies Eq. (5.28) and the solution representing a transmitted wave of intensity $|A_T|^2$ is

$$\psi_E(x) = A_T\, e^{+ikx}. \tag{5.34}$$

If we smoothly join these solutions together at $x = 0$ and at $x = a$, we can find the intensities of the reflected and transmitted waves. In particular, we can derive expressions for the ratios

$$R = \frac{|A_R|^2}{|A_I|^2} \quad \text{and} \quad T = \frac{|A_T|^2}{|A_I|^2}. \tag{5.35}$$

Because the probability of finding a particle at x is proportional to $|\psi_E(x)|^2$, these ratios are probabilities: R is the probability that the particle is reflected and T is the probability that the particle is transmitted, and the sum of these probabilities is equal to one, i.e.

$$R + T = 1. \tag{5.36}$$

This interpretation of R and T is confirmed in problem 5 at the end of this chapter.

We shall confine our attention to a particle with an energy below the barrier and derive an approximate expression for the transmission probability which is useful when the barrier is wide. In other words, we will find the probability of tunnelling through a wide barrier.

Tunnelling through wide barriers

The eigenfunction of a particle with energy below the barrier is given by

$$\psi_E(x) = \begin{cases} A_I\, e^{+ikx} + A_R\, e^{-ikx} & \text{if } -\infty < x < 0 \\ B\, e^{-\beta x} + B'\, e^{+\beta x} & \text{if } 0 < x < a \\ A_T\, e^{+ikx} & \text{if } a < x < \infty, \end{cases} \tag{5.37}$$

with constants A_I, A_R, A_T, B and B' having values which ensure that ψ_E and $d\psi_E/dx$ are continuous at $x = 0$ and at $x = a$. Continuity at $x = 0$ requires

$$A_I + A_R = B + B' \quad \text{and} \quad ikA_I - ikA_R = -\beta B + \beta B',$$

and continuity at $x = a$ requires

$$Be^{-\beta a} + B'e^{+\beta a} = A_T e^{ika} \quad \text{and} \quad -\beta Be^{-\beta a} + \beta B'e^{+\beta a} = ikA_T e^{ika}.$$

Our objective is to find an expression for the tunnelling probability. To do so we shall express the constants A_T and A_I in terms of B. First, we use continuity at $x = 0$ to give

$$2ikA_I = -(\beta - ik)B + (\beta + ik)B'. \tag{5.38}$$

Second, we use continuity at $x = a$ to give

$$A_T e^{ika} = \frac{2\beta}{(\beta - ik)} Be^{-\beta a} \quad \text{and} \quad B' = Be^{-2\beta a} \frac{(\beta + ik)}{(\beta - ik)}. \tag{5.39}$$

It is now straightforward to express both A_I and A_T in terms of B and obtain the tunnelling probability $T = |A_T|^2/|A_I|^2$.

To avoid algebraic tedium, we shall simplify matters by assuming that the particle encounters a wide barrier with $\exp(-2\beta a) << 1$. When this is the case, the constant B' is much smaller than B and we can approximate Eq. (5.38) by

$$2ikA_I \approx -(\beta - ik)B.$$

If we combine this equation with Eq. (5.39), we obtain

$$A_T e^{ika} \approx -\frac{4ik\beta e^{-\beta a}}{(\beta - ik)^2} A_I.$$

This implies that the tunnelling probability is approximately given by

$$T \approx \left[\frac{16k^2\beta^2}{(\beta^2 + k^2)^2} \right] e^{-2\beta a}.$$

If we rewrite this equation using

$$k = \frac{\sqrt{2mE}}{\hbar} \quad \text{and} \quad \beta = \frac{\sqrt{2m(V_B - E)}}{\hbar},$$

we obtain

$$T \approx \left[\frac{16E(V_B - E)}{V_B^2} \right] e^{-2\beta a}. \tag{5.40}$$

This approximate expression for the tunnelling probability is valid when $e^{-2\beta a} << 1$. When this is the case, the wave function inside the barrier is dominated by the exponentially decaying term $Be^{-\beta x}$ and the probability of tunnelling a barrier of width a is proportional to $e^{-2\beta a}$. In fact, this exponential dependence on the penetration parameter β and on the width of the barrier a is the most important, and the most useful, feature of the tunnelling formula Eq. (5.40).

We shall illustrate the importance of barrier penetration by considering two processes, one involving tunnelling electrons and the other tunnelling protons.

Tunnelling electrons

We know from the photoelectric effect that the minimum energy needed to eject an electron from the surface of a metal is of the order of a few electron volts. This energy is needed because electrons in a metal reside in an attractive potential energy field which increases at the surface of the metal to give a potential step which is a few electron volts above the energy of the most energetic electrons in the metal.

When two metal surfaces are placed in close proximity, there are two regions of low potential energy separated by a potential barrier which is similar to that shown in Fig. 5.4. But this barrier does not prevent electrons from moving across the gap between the surfaces. Electrons are quantum particles that can tunnel through the barrier with a probability given by Eq. (5.40), or approximately by

$$T \approx e^{-2\beta a} \quad \text{with} \quad \beta = \frac{\sqrt{2m_e(V_B - E)}}{\hbar}. \tag{5.41}$$

In this expression a is the width of the gap between the metal surfaces, m_e is the mass of the electron, V_B is the barrier height at the surface of the metal and E is the energy of the most energetic electrons in the metal.

Equation (5.41) implies that the tunnelling probability is a very sensitive function of the gap. Indeed, if the gap a changes by a small amount Δa, the fractional change in the tunnelling probability is given by

$$\frac{\Delta T}{T} = -2\beta \, \Delta a. \tag{5.42}$$

To illustrate the numerical significance of this formula, we shall consider a typical situation in which 4 eV are needed to eject an electron from the metal. This implies that the height of the barrier V_B is about 4 eV higher than the electron energy E and that the penetration parameter is given by

$$\beta = \frac{\sqrt{2m_e(V_B - E)}}{\hbar} \approx 10^{10} \, \text{m}^{-1}. \tag{5.43}$$

By substituting this value for β into Eq. (5.42) we can illustrate the incredible sensitivity of the electron tunnelling probability. For example, there is a measurable 2% change in the probability when the gap between the surfaces changes by a mere 0.001 nm!

The extreme sensitivity of electron tunnelling is exploited in a device called the *scanning tunnelling microscope*. In this device a sharp metal probe is positioned near to a surface under investigation. The separation is made small enough to induce the tunnelling of electrons between the probe and the surface, and a potential difference between the probe and the surface is also established so that there is a net current of electrons in one direction. As the probe is moved or scanned across the surface, surface features of atomic dimensions will give rise to measurable changes in the current of tunnelling electrons. In this way the scanning tunnelling microscope can produce a map of the locations of individual atoms on the surface.

Tunnelling protons

The centre of the sun consists of an ionized gas of electrons, protons and light atomic nuclei at a temperature T of about 10^7 K. The protons and other light nuclei collide frequently, occasionally get close and occasionally fuse to release energy which is ultimately radiated from the solar surface as sunshine.

To understand the issues involved in the generation of solar thermonuclear energy, we shall focus on two protons approaching each other near the centre of the sun. They move in the ionized gas with thermal kinetic energies of the order of

$$E \approx kT \approx 1\,\text{keV}.$$

The mutual potential energy of the two protons depends on their separation. As illustrated in Fig. 5.5, the potential energy at large separation r is dominated by the repulsive Coulomb potential

$$V(r) = \frac{e^2}{4\pi\epsilon_0 r}.$$

But at small separations, when r becomes comparable with the range of nuclear forces given by $r_N \approx 2 \times 10^{-15}$ m, the potential energy becomes attractive. The net effect is a Coulomb barrier which rises to a height of about 1 MeV at a separation of about 2×10^{-15} m or 2 fm.

Thus, when protons approach each other near the centre of the sun, they do so with energies of the order of keV and they encounter a Coulomb barrier measured in MeV. According to classical physics, there is a well-defined distance of closest approach r_C, which is given by

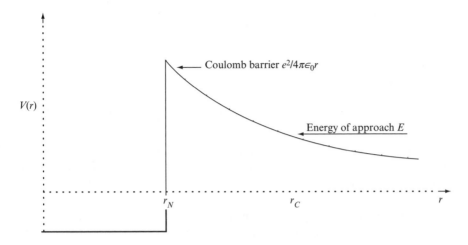

Fig. 5.5 A schematic representation of the potential energy $V(r)$ of two protons with separation r. When the separation is less than $r_N \approx 2 \times 10^{-15}$ m, there is a strong nuclear attraction and the protons may fuse to form a deuteron. Nuclear fusion is hindered by a Coulomb barrier which is approximately 1 MeV high. The distance r_C is the classical distance of closest approach for protons with an energy of approach equal to E.

$$E = \frac{e^2}{4\pi\epsilon_0 r_C}. \qquad (5.44)$$

Because this distance is three orders of magnitude larger that the range of nuclear forces r_N, the prospect of a close encounter and the possibility of nuclear fusion look dim. Indeed, at first sight, the sun is not hot enough to shine.

In fact, thermonuclear fusion in the sun, and in other stars, is only possible because protons are quantum particles that can tunnel through Coulomb barriers. Two protons with an energy of approach E are described by an eigenfunction $\psi(\mathbf{r})$ which obeys the time-independent Schrödinger equation,

$$\left[-\frac{\hbar^2}{2\mu}\nabla^2 + V(r) \right]\psi(\mathbf{r}) = E\psi(\mathbf{r}), \qquad (5.45)$$

where $V(r)$ is the potential shown in Fig. 5.5 and μ is equal to $m_p/2$, the reduced mass of two protons.[4] For two protons with low energy, the relevant eigenfunction has no angular dependence and it has the form

---

[4] The reduced mass $m_p/2$ is used because the kinetic energy of two protons approaching each other with equal and opposite momentum p is given by

$$E = \frac{p^2}{2m_p} + \frac{p^2}{2m_p} = \frac{p^2}{m_p}.$$

$$\psi(\mathbf{r}) = \frac{u(r)}{r},$$

(5.46)

where $u(r)$, as shown in problem 2 at the end of the chapter, obeys

$$-\frac{\hbar^2}{2\mu}\frac{d^2u}{dr^2} + V(r)u = E.$$

(5.47)

We shall not attempt to solve this equation. Instead, we shall use the results we obtained for a one-dimensional barrier to write down a plausible form of the eigenfunction and then estimate the probability of tunnelling through a Coulomb barrier.

We begin by considering a three-dimensional barrier with constant height V_B and width $r_C - r_N$. In this case, the function $u(r)$ decays exponentially in the classically forbidden region as r gets smaller and it is given by

$$u(r) \propto e^{\beta r},$$

where β is given by

$$E = -\frac{\hbar^2\beta^2}{2\mu} + V_B.$$

The probability that the protons tunnel from $r = r_C$ to $r = r_N$ is approximately equal to the ratio of $|u(r_N)|^2$ to $|u(r_C)|^2$, and it is given by

$$T \approx |\exp[-\beta(r_C - r_N)]|^2.$$

(5.48)

We now consider the Coulomb barrier shown in Fig. 5.5. In this case, the eigenfunction in the classically forbidden region ($r_N < r < r_C$) is again approximately given by

$$u(r) \propto e^{\beta r},$$

but β now depends on r because it is given by

$$E = -\frac{\hbar^2\beta^2}{2\mu} + \frac{e^2}{4\pi\epsilon_0 r}.$$

The probability that the two protons tunnel from $r = r_C$ to $r = r_N$ is now approximately given by a generalization of Eq. (5.48)

$$T \approx \left| \exp\left[-\int_{r_N}^{r_C} \beta \, dr \right] \right|^2 . \tag{5.49}$$

If we assume that $r_C \gg r_N$ and evaluate the integral in Eq. (5.49) by substituting $r = r_C \cos^2 \theta$, we find that

$$T \approx \exp\left[-\left(\frac{E_G}{E} \right)^{1/2} \right] \tag{5.50}$$

where E is the relative energy of the protons and E_G is defined by

$$E_G = \left(\frac{e^2}{4\pi\epsilon_0 \hbar c} \right)^2 2\pi^2 \mu c^2 . \tag{5.51}$$

The energy E_G is called *the Gamow energy* and its value is 493 keV.

We can now estimate the probability that two protons tunnel through the Coulomb barrier which normally keeps them well apart when they collide near the centre of the sun. By substituting a typical thermal energy of $E = 1$ keV into Eq. (5.50), we obtain

$$T \approx \exp[-22] \approx 3 \times 10^{-10} .$$

Thus, with a probability of about one in 3 billion, protons colliding near the centre of the sun tunnel through the Coulomb barrier. And when they do so they have a chance of fusing and releasing thermonuclear energy.

In practice, stars evolve slowly by adjusting their temperature so that the average thermal energy of nuclei is well below the Coulomb barrier. Fusion then proceeds at a rate proportional to the tunnelling probability. Because this probability is very low, fusion proceeds at a slow pace and the nuclear fuel lasts for an astronomically long time scale.

PROBLEMS 5

1. Consider a particle of mass m in the one-dimensional potential energy field

$$V(x) = \begin{cases} 0 & \text{if } -\infty < x < -a \\ -V_0 & \text{if } -a < x < +a \\ 0 & \text{if } +a < x < +\infty. \end{cases}$$

Because the potential is symmetric about $x = 0$, there are two types of energy eigenfunctions. There are symmetric eigenfunctions which obey

$$\psi(x) = +\psi(-x)$$

and antisymmetric eigenfunctions which obey

$$\psi(x) = -\psi(-x).$$

Symmetric eigenfunctions are said to have positive parity and antisymmetric eigenfunctions are said to have negative parity.

(a) Show, by considering the energy eigenvalue equation in the three regions of x, that a positive parity eigenfunction with energy $E = -\hbar^2\alpha^2/2m$ has the form:

$$\psi(x) = \begin{cases} A e^{+\alpha x} & \text{if } -\infty < x < -a \\ C \cos k_0 x & \text{if } -a < x < +a \\ A e^{-\alpha x} & \text{if } +a < x < +\infty, \end{cases}$$

where A and C are constants and $k_0 = \sqrt{2m(E + V_0)/\hbar^2}$.

(b) Show that the continuity of $\psi(x)$ and $d\psi/dx$ at the edges of the potential well implies that

$$\alpha = k_0 \tan k_0 a.$$

(c) By seeking a graphical solution of the equations

$$\alpha = k_0 \tan k_0 a \quad \text{and} \quad \alpha^2 + k_0^2 = w^2,$$

where $w = \sqrt{2mV_0/\hbar^2}$, show that there is one bound state if

$$0 < w < \frac{\pi}{2a},$$

two bound states if

$$\frac{\pi}{2a} < w < \frac{3\pi}{2a},$$

and so on.

(d) Now consider a negative-parity eigenfunction and confirm that the bound state energies are identical to those given by the potential illustrated in Fig. 5.1.

2. A particle of mass m moves in three dimensions in a potential energy field

$$V(r) = \begin{cases} -V_0 & \text{if } r < R \\ 0 & \text{if } r > R, \end{cases}$$

where r is the distance from the origin. Its energy eigenfunctions $\psi(\mathbf{r})$ are governed by

$$-\frac{\hbar^2}{2m}\nabla^2\psi + V(r)\psi = E\psi,$$

where, in spherical polar coordinates,

$$\nabla^2\psi = \frac{1}{r}\frac{\partial^2(r\psi)}{\partial r^2} + \frac{1}{r^2}\left[\frac{\partial^2\psi}{\partial\theta^2} + \frac{\cos\theta}{\sin\theta}\frac{\partial\psi}{\partial\theta} + \frac{1}{\sin^2\theta}\frac{\partial^2\psi}{\partial\phi^2}\right].$$

Consider spherically symmetric eigenfunctions with no angular dependence of the form

$$\psi(r) = \frac{u(r)}{r}.$$

(a) Show that

$$-\frac{\hbar^2}{2m}\frac{d^2u}{dr^2} + V(r)u = E.$$

(b) By solving for $u(r)$ in the regions $r < R$ and $r > R$ and by imposing appropriate boundary conditions, show that an eigenfunction of a bound state with energy $E = -\hbar^2\alpha^2/2m$ has the form

$$\psi(r) = \begin{cases} C\sin k_0 r/r & \text{if } r < R \\ A e^{-\alpha r}/r & \text{if } r > R, \end{cases}$$

where C and A are constants and $k_0 = \sqrt{2m(V_0 + E)/\hbar^2}$.

(c) Show that there is one bound state of this kind if the depth of the well obeys

$$\frac{\hbar^2\pi^2}{8mR^2} < V_0 < \frac{9\hbar^2\pi^2}{8mR^2}.$$

(Many of the steps in this question are almost identical to those carried out in Section 5.1 for the one-dimensional potential given in Fig 5.1.)

3. A particle with energy $E = \hbar^2\pi^2/m$ is scattered by the potential well shown in Fig. 5.1 with depth $V_0 = 2\hbar^2\pi^2/ma^2$. Use Eq. (5.19) to show that the eigenfunction has a phase shift, in radians, which can be taken as

$$\delta(E) = 3.53.$$

[Note that any integer multiple of π can be added to the phase shift satisfying Eq. (5.19).]

4. (a) A particle in a potential $V(x)$ has definite energy

$$E = -\frac{\hbar^2\alpha^2}{2m}$$

and an eigenfunction

$$\psi(x) = \begin{cases} Nxe^{-\alpha x} & \text{if } 0 \leq x < \infty \\ 0 & \text{elsewhere,} \end{cases}$$

where N and α are positive real constants. Verify that the potential is given by

$$V(x) = \begin{cases} -\alpha\hbar^2/mx & \text{if } 0 \leq x < \infty \\ \infty & \text{elsewhere.} \end{cases}$$

(b) Given that the eigenfunction $\psi(x)$ given in part (a) describes the ground state of the particle in the potential $V(x)$, roughly sketch the eigenfunction which describes the first excited state.

5. In problem 8 at the end of Chapter 3, we showed that the function $j(x, t)$, defined by

$$j(x, t) = \frac{i\hbar}{2m}\left[\Psi^*\frac{\partial\Psi}{\partial x} - \Psi\frac{\partial\Psi^*}{\partial x}\right],$$

can be interpreted as a current of probability because

$$j(x_2, t) - j(x_1, t) = \frac{d}{dt}\left[\int_{x_1}^{x_2}|\Psi(x, t)|^2\,dx\right].$$

Consider a particle with a stationary state wave function

$$\Psi(x,\ t) = \psi(x)\,e^{-iEt/\hbar}$$

incident on the barrier illustrated in Fig. 5.4. Assume that on the left of the barrier

$$\psi_E(x) = A_I\,e^{+ikx} + A_R\,e^{-ikx}$$

and on the right of the barrier

$$\psi_E(x) = A_T\,e^{+ikx}.$$

(a) Show that on the left

$$j(x,\ t) = |A_I|^2\frac{\hbar k}{m} - |A_R|^2\frac{\hbar k}{m}$$

and that on the right

$$j(x,\ t) = |A_T|^2\frac{\hbar k}{m}.$$

(b) By noting that the position probability density $|\Psi(x,\ t)|^2$ is constant for a stationary state, show that

$$|A_I|^2\frac{\hbar k}{m} = |A_R|^2\frac{\hbar k}{m} + |A_T|^2\frac{\hbar k}{m}.$$

(This implies that the incident probability current equals the sum of the reflected and transmitted probability currents, and that the reflection and transmitted probabilities satisfy the relation

$$R + T = 1.$$

As expected, the sum of the probabilities of the only possible two outcomes of the encounter, reflection and transmission, is unity.)

6. Consider a particle with an energy E above the square potential barrier illustrated in Fig. 5.4. As in Section 5.2, define wave numbers k_B and k by

$$E = \frac{\hbar^2 k_B^2}{2m} + V_B = \frac{\hbar^2 k^2}{2m}$$

and show that the transmission probability is given by

$$T = \frac{|A_T|^2}{|A_I|^2} = \frac{1}{1 + (S\sin 2k_B a)^2}$$

and that the reflection probability is given by

$$R = \frac{|A_R|^2}{|A_I|^2} = \frac{(S\sin 2k_B a)^2}{1 + (S\sin 2k_B a)^2},$$

where

$$S = \frac{(k^2 - k_B^2)}{2kk_B}.$$

Show that the barrier is completely transparent for certain values of the energy.

7. Find the classical distance of closest approach for two protons with an energy of approach equal to 2 keV. Estimate the probability that the protons penetrate the Coulomb barrier tending to keep them apart. Compare this probability with the corresponding probability for two ^{4}He nuclei with the same energy of approach.

8. We have seen that the tunnelling through a Coulomb barrier plays a crucial role in thermonuclear fusion. It also plays a crucial role in the alpha-decay of nuclei such as ^{235}U. In the simplest model for alpha-decay, the alpha-particle is preformed and trapped within the nucleus by a potential similar to that shown in Fig. 5.5. The mean rate of decay, λ, is then the product of the frequency v with which the alpha-particle hits the confining barrier, multiplied by the probability of penetration of the Coulomb barrier; this probability is given by Eq. (5.50).

   Write down an approximate expression for the decay rate in terms of v, E_G and the energy released by alpha-decay, E. The half-life for the alpha-decay of ^{235}U is 7.1×10^8 years and the energy released is $E = 4.68$ MeV. Estimate the half-life for the alpha-decay of ^{239}Pu given that the energy released in this decay is 5.24 MeV.

6

The harmonic oscillator

The harmonic oscillator played a leading role in the development of quantum mechanics. In 1900, Planck made the bold assumption that atoms acted like oscillators with quantized energy when they emitted and absorbed radiation; in 1905, Einstein assumed that electromagnetic radiation acted like electromagnetic harmonic oscillators with quantized energy; and in 1907, Einstein assumed that the elastic vibrations of a solid behaved as a system of mechanical oscillators with quantized energy. These assumptions were invoked to account for black body radiation, the photoelectric effect and the temperature dependence of the specific heats of solids. Subsequently, quantum theory provided a fundamental description of both electromagnetic and mechanical harmonic oscillators.

This chapter deals with the quantum mechanical behaviour of a particle in a harmonic oscillator potential. We shall find the energy eigenvalues and eigenfunctions, and explore the properties of stationary and non-stationary quantum states. These states are very important in molecular and solid state physics, nuclear physics and, more generally, in quantum field theory.

6.1 THE CLASSICAL OSCILLATOR

The simplest example of a harmonic oscillator is a particle on a spring with elastic constant k. When the particle is displaced from equilibrium by a distance x, there is a force $F = -kx$ which opposes the displacement. Because the work needed to move the particle from x to $x + \mathrm{d}x$ is $kx\,\mathrm{d}x$, the potential energy stored by displacing the particle by a finite distance x is

$$V(x) = \int_0^x kx'\,\mathrm{d}x' = \frac{1}{2}kx^2. \tag{6.1}$$

This potential energy is converted into kinetic energy when the particle is released.

The equation of motion for a particle of mass m is

$$m\frac{d^2x}{dt^2} = -kx,$$ (6.2)

and this is usually rewritten as

$$\ddot{x} = -\omega^2 x,$$ (6.3)

where $\ddot{x}$ is the acceleration of the particle and $\omega = \sqrt{k/m}$. The general solution is

$$x = A\cos(\omega t + \alpha),$$ (6.4)

where A and α are two constants which may be determined by specifying the initial position and velocity of the particle; for example, if the particle is released from rest at position x_0, then $A = x_0$ and $\alpha = 0$.

Eq. (6.4) describes simple harmonic motion with amplitude A, phase α and angular frequency ω or period $2\pi/\omega$. During the motion, the potential energy rises and falls as the kinetic energy falls and rises. But the total energy E, the sum of the kinetic and potential energies, remains constant and equal to

$$E = \tfrac{1}{2}m\dot{x}^2 + \tfrac{1}{2}kx^2 = \tfrac{1}{2}m\omega^2 A^2.$$ (6.5)

In the real world of quantum mechanics, simple harmonic motion with definite energy, frequency, phase and amplitude never really happens. We shall see that oscillators either have definite energy and do not oscillate, or they oscillate with uncertain energy. However, we shall see that, in special circumstances, the oscillations are almost like simple harmonic motion.

6.2   THE QUANTUM OSCILLATOR

The defining property of a quantum system is its Hamiltonian operator. For a one-dimensional harmonic oscillator the Hamiltonian operator is

$$\hat{H} = \frac{\hat{p}^2}{2m} + \frac{1}{2}m\omega^2\hat{x}^2,$$ (6.6)

or, if we use Eq. (3.30),

$$\hat{H} = -\frac{\hbar^2}{2m}\frac{\partial^2}{\partial x^2} + \frac{1}{2}m\omega^2 x^2.$$ (6.7)

The first term represents the kinetic energy operator for a particle of mass m and the second term represents the potential energy operator for a particle in a potential well which, in classical physics, would give rise to simple harmonic motion with angular frequency ω.

The behaviour of a particle in a harmonic oscillator potential is more varied in quantum physics than in classical physics. There are an infinite number of quantum states; some are stationary states with definite energy and some are non-stationary states with uncertain energy. Each of these states is described by a wave function $\Psi(x, t)$ which satisfies the Schrödinger equation,

$$i\hbar \frac{\partial \Psi}{\partial t} = \hat{H}\Psi, \tag{6.8}$$

with the Hamiltonian $\hat{H}$ given by Eq. (6.7).

The quantum states with definite energy will be our first concern. As we saw in Section 4.3, states with energy E are represented by a wave function of the form

$$\Psi(x, t) = \psi(x)\,e^{-iEt/\hbar}, \tag{6.9}$$

where $\psi(x)$ is an eigenfunction belonging to an energy eigenvalue E. If we substitute for $\Psi(x, t)$ into Eq. (6.8) and use Eq. (6.7), we obtain the energy eigenvalue equation:

$$\left[-\frac{\hbar^2}{2m}\frac{d^2}{dx^2} + \frac{1}{2}m\omega^2 x^2 \right]\psi(x) = E\psi(x). \tag{6.10}$$

When we seek solutions to this equation, we shall impose the physical requirement that the wave function of the particle is normalizable. To do this, we shall require the eigenfunctions to go to zero at infinity; i.e.

$$\psi(x) \to 0 \quad \text{as} \quad x \to \pm\infty. \tag{6.11}$$

Because the sides of the harmonic oscillator potential, like the walls of the infinite square-well potential, are infinitely high, we expect an infinite number of quantized eigenvalues. We shall denote these by E_n, and the corresponding eigenfunctions by $\psi_n(x)$, where n is a quantum number. We shall follow the convention of labelling the ground state by $n = 0$ and the first, second and third excited states, etc, by $n = 1, 2, 3, \ldots$.[1]. Because we wish to emphasise the physics of the harmonic oscillator, we shall defer the mathematical problem of finding the energy eigenvalues and eigenfunctions until Section 6.6.

---

[1] Note, the convention for the infinite square-well potential is different. The ground state for this potential has the label $n = 1$; see Eq. (4.38).

6.3   QUANTUM STATES

We shall first describe the properties of the stationary states of the harmonic oscillator and then explore the extent to which the non-stationary states resemble classical simple harmonic motion.

Stationary states

In Section 6.6 we shall show that the energy eigenvalues of a harmonic oscillator with classical angular frequency ω are given by

$$E_n = (n + \tfrac{1}{2})\, \hbar\omega, \quad \text{with} \quad n = 0, 1, 2, 3, \ldots\ldots \tag{6.12}$$

As illustrated in Fig. 6.1, the energy levels have equal spacing $\hbar\omega$ and the lowest energy level is $E_0 = \frac{1}{2}\hbar\omega$.

The wave function of a particle with energy E_n has the form

$$\Psi(x, t) = \psi_n(x)\, e^{-iE_n t/\hbar} \tag{6.13}$$

where $\psi_n(x)$ is the energy eigenfunction. The eigenfunctions of the four lowest states of the oscillator are given in Table 6.1 and they are shown in Fig. 6.2. We note that the nth eigenfunction has n nodes; i.e. there are n values of x for which $\psi_n(x) = 0$. In Section 4.4 we showed that the eigenfunctions of an infinite square well have a similar property. In general, eigenfunctions of excited bound states always have a number of nodes which increases with the degree of excitation, just as classical normal modes of oscillation of higher frequency always have a higher number of nodes.

TABLE 6.1 Normalized eigenfunctions for the four lowest states of a one-dimensional harmonic oscillator. The parameter which determines the spatial extent of the eigenfunctions is $a = \sqrt{\hbar/m\omega}$

Quantum number	Energy eigenvalue	Energy eigenfunction
$n = 0$	$E_0 = \dfrac{1}{2}\,\hbar\omega$	$\psi_0(x) = \left(\dfrac{1}{a\sqrt{\pi}}\right)^{\frac{1}{2}} e^{-x^2/2a^2}$
$n = 1$	$E_1 = \dfrac{3}{2}\,\hbar\omega$	$\psi_1(x) = \left(\dfrac{1}{2a\sqrt{\pi}}\right)^{\frac{1}{2}} 2\left(\dfrac{x}{a}\right) e^{-x^2/2a^2}$
$n = 2$	$E_2 = \dfrac{5}{2}\,\hbar\omega$	$\psi_2(x) = \left(\dfrac{1}{8a\sqrt{\pi}}\right)^{\frac{1}{2}} \left[2 - 4\left(\dfrac{x}{a}\right)^2\right] e^{-x^2/2a^2}$
$n = 3$	$E_3 = \dfrac{7}{2}\,\hbar\omega$	$\psi_3(x) = \left(\dfrac{1}{48a\sqrt{\pi}}\right)^{\frac{1}{2}} \left[12\left(\dfrac{x}{a}\right) - 8\left(\dfrac{x}{a}\right)^3\right] e^{-x^2/2a^2}$

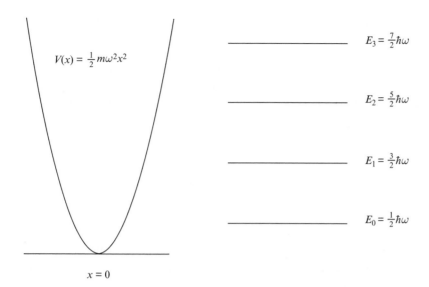

Fig. 6.1   Low-lying energy levels of a particle of mass m confined by a one-dimensional harmonic oscillator potential.

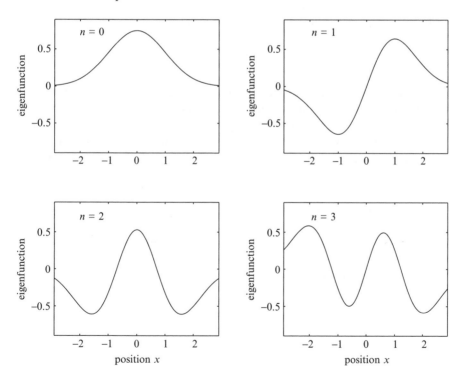

Fig. 6.2   The spatial shapes of the eigenfunctions $\psi_n(x)$ for the four lowest states of a one-dimensional harmonic oscillator with length parameter $a = \sqrt{\hbar/m\omega} = 1$.

The observable properties of a quantum state with energy E_n include the following:

- The eigenfunction has an observable property called *parity*, which will be discussed further in Sections 9.1 and 9.4. If the position coordinate is changed from x to $-x$, the eigenfunction has a definite symmetry:

$$\psi_n(-x) = +\psi_n(x) \qquad \text{if } n \text{ is even,}$$

and

$$\psi_n(-x) = -\psi_n(x) \qquad \text{if } n \text{ is odd.}$$

Oscillator states with even n are said to have *positive parity* and oscillator states with odd n are said to have *negative parity*. This property arises because the Hamiltonian of the harmonic oscillator, Eq. (6.7), is unchanged under the transformation

$$x \to x' = -x.$$

- The position probability density of the particle,

$$|\Psi_n(x, t)|^2 = |\psi_n(x)|^2, \qquad (6.14)$$

is time-independent, as expected from Section 4.8, where we showed that quantum states of definite energy are stationary states, i.e. states with no observable time dependence. As illustrated in Fig. 6.3, the particle can have any location between $x = -\infty$ and $x = +\infty$, in marked contrast with a classical particle which is confined to the region $-A < x < +A$, where A is the amplitude of oscillation.

- The position expectation values are[2]

$$\langle x \rangle = 0 \quad \text{and} \quad \langle x^2 \rangle = \left(n + \tfrac{1}{2}\right)a^2,$$

so that the uncertainty in position is

$$\Delta x = \sqrt{\langle x^2 \rangle - \langle x \rangle^2} = \sqrt{\left(n + \tfrac{1}{2}\right)}a,$$

where $a = \sqrt{\hbar/m\omega}$.

---

[2] The expectation values of position and momentum can derived using the results of problems 10 and 11 at the end of this chapter.

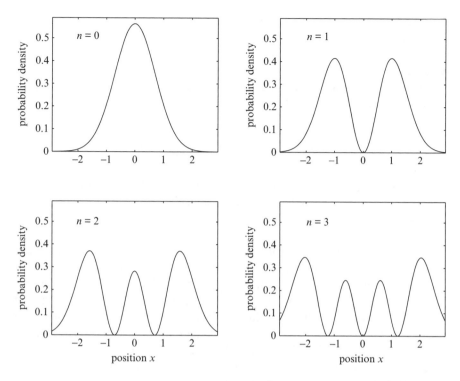

Fig. 6.3 The position probability densities $|\psi_n(x)|^2$ for the four lowest states of a one-dimensional harmonic oscillator with length parameter $a = \sqrt{\hbar/m\omega} = 1$.

- The momentum expectation values are

$$\langle p \rangle = 0 \quad \text{and} \quad \langle p^2 \rangle = \left(n + \tfrac{1}{2}\right) \frac{\hbar^2}{a^2},$$

so that the uncertainty in the particle's momentum is

$$\Delta p = \sqrt{\langle p^2 \rangle - \langle p \rangle^2} = \sqrt{\left(n + \tfrac{1}{2}\right)} \frac{\hbar}{a}.$$

- The product of the position and momentum uncertainties is

$$\Delta x \, \Delta p = \left(n + \tfrac{1}{2}\right)\hbar,$$

which agrees with the Heisenberg uncertainty principle, Eq. (1.15), which asserts that in general

$$\Delta x \, \Delta p \geq \tfrac{1}{2}\hbar.$$

We note that when $n = 0$, i.e. when the particle is in the ground state of the oscillator, the product of Δx and Δp is as small as it can be.

- Because the position and momentum of the particle are uncertain, the potential energy and the kinetic energy are uncertain. The expectation values of these uncertain observables are

$$\frac{1}{2}m\omega^2\langle x^2\rangle = \frac{1}{2}E_n \quad \text{and} \quad \frac{\langle p^2\rangle}{2m} = \frac{1}{2}E_n.$$

Not surprisingly, the sum of the expectation values of the uncertain potential and kinetic energies is equal to E_n, the sharply defined total energy of the state.

Finally, it is useful to consider in general terms why the quantum ground state of the harmonic oscillator is so different from the classical ground state in which the particle lies at rest at the bottom of the well with zero kinetic energy and zero potential energy. When a quantum particle is precisely localized at the centre of the well, it has a highly uncertain momentum and, hence, a high kinetic energy. Similarly, when its momentum is precisely zero, it has a highly uncertain position and it may be found in regions of high potential energy. It follows that the sum of the kinetic and potential energies of a quantum particle in a harmonic oscillator potential has a minimum value when its position and momentum are uncertain, but not too uncertain. This minimum is called the *zero point energy of the harmonic oscillator*. A lower bound for this energy is derived using these ideas in problem 1 at the end of this chapter.

Non-stationary states

The general wave function of a particle in a harmonic oscillator potential has the form

$$\Psi(x, t) = \sum_{n=0, 1, 2...} c_n \psi_n(x) \, e^{-iE_n t/\hbar}. \tag{6.15}$$

This wave function represents a state of uncertain energy because when the energy is measured many outcomes are possible: $E_0 = \frac{1}{2}\hbar\omega$, $E_1 = \frac{3}{2}\hbar\omega$, ... with probabilities $|c_0|^2, |c_1|^2, \ldots$.

This wave function also represents a non-stationary state, a state with time-dependent observable properties. For example, the position probability amplitude $|\Psi(x, t)|^2$ has time-dependent terms which arise from the interference of terms involving different energy eigenfunctions, $\psi_n(x)$. In particular, the interference of $\psi_m(x)\,\mathrm{e}^{-iE_m t/\hbar}$ with $\psi_n(x)\,\mathrm{e}^{-iE_n t/\hbar}$ gives rise to a term which oscillates with angular frequency

$$\omega_{m,n} = \frac{|E_m - E_n|}{\hbar},$$

(6.16)

which is an integer multiple of the classical angular frequency ω. Thus, the position probability density, $|\Psi(x,t)|^2$, can oscillate with a range of angular frequencies: $\omega, 2\omega, 3\omega$, etc.

At first sight, we expect the average position of the particle to oscillate with the same frequencies, because, when we evaluate the expectation value

$$\langle x(t) \rangle = \int_{-\infty}^{+\infty} \Psi^*(x,t)x\Psi(x,t)\,\mathrm{d}x,$$

we obtain terms like

$$c_m^* c_n x_{m,n}\,\mathrm{e}^{i\omega_{m,n}t}, \quad \text{where} \quad x_{m,n} = \int_{-\infty}^{+\infty} \psi_m^*(x)x\psi_n(x)\,\mathrm{d}x.$$

But most of these terms are zero because, as shown in problem 11, the integral $x_{m,n}$ is zero when $|m - n|$ is greater than one. It follows that the average position only oscillates with angular frequency ω, and so does the average momentum. In general, the average position and momentum of a particle of mass m with wave function (6.15) are given by

$$\langle x(t) \rangle = A\cos(\omega t + \alpha) \quad \text{and} \quad \langle p(t) \rangle = -m\omega A \sin(\omega t + \alpha).$$

(6.17)

Thus, to some extent, the particle oscillates back and forth like a classical oscillator. But the resemblance to classical simple harmonic motion may not be close because the uncertainties in position and momentum may be large and they may change as the particle oscillates back and forth. Accordingly, it is instructive to identify quasi-classical states in which the motion most closely resembles classical simple harmonic motion.

It can be shown that the wave functions of these quasi-classical states are given by Eq. (6.15) with energy probability amplitudes c_n which satisfy the Poisson probability distribution[3]

$$|c_n|^2 = \frac{\bar{n}^n}{n!}\,\mathrm{e}^{-\bar{n}}, \quad \text{with} \quad \bar{n} \gg 1.$$

---

[3] Quasi-classical states are fully described in *Quantum Mechanics*, vol. I, C Cohen-Tannoudji, B Diu and F Laloë, John Wiley & Sons (1977).

In problem 1 at the end of Chapter 3 we showed that the mean and standard deviation of this distribution are $\bar{n}$ and $\sqrt{\bar{n}}$. It follows that the mean and standard deviation of the energy of a quasi-classical state is given by

$$\langle E \rangle = (\bar{n} + \tfrac{1}{2})\hbar\omega \quad \text{and} \quad \Delta E = \sqrt{\bar{n}}\hbar\omega.$$

When the average excitation $\bar{n}$ is high, the relative uncertainty in energy is given by

$$\frac{\Delta E}{\langle E \rangle} = \frac{\sqrt{\bar{n}}}{\bar{n} + \tfrac{1}{2}} \to \frac{1}{\sqrt{\bar{n}}}$$

so that the uncertainty in the energy becomes less important. Because the uncertainties in the position and momentum also become less important, the motion of the particle approaches the impossible perfection of classical simple harmonic motion.

At the turn of the twentieth century, Planck and Einstein made the inspired guess that oscillators which exhibit simple harmonic motion could also have quantized energies. We have followed a logical path in the opposite direction and indicated how quantum oscillators, which have quantized energies, can almost exhibit simple harmonic motion.

6.4   DIATOMIC MOLECULES

To a first approximation a diatomic molecule consists of two nuclei held together in an effective potential which arises from the Coulomb interactions of the electrons and nuclei and the quantum behaviour of the electrons. This effective potential determines the strength of the molecular bond between the nuclei and it also governs the vibrational motion of the nuclei. The effective potential and the vibrational energy levels of the simplest diatomic molecule, the hydrogen molecule, are illustrated in Fig. 6.4.

We note from Fig. 6.4 that an effective potential for a diatomic molecule, $V_e(r)$, has a minimum and that near this minimum the shape is like a harmonic oscillator potential. Indeed, if r_0 denotes the separation at which the effective potential has a minimum and $x = r - r_0$ denotes a small displacement from r_0, we can write

$$V_e(r) \approx \tfrac{1}{2}kx^2$$

where k is a constant. This equation implies that when the nuclei are displaced a distance x from their equilibrium separation of r_0, there is a restoring force of magnitude kx and the potential energy increases by $\tfrac{1}{2}kx^2$. The constant k is an effective elastic constant which characterizes the strength of the molecular bond between the nuclei in the molecule.

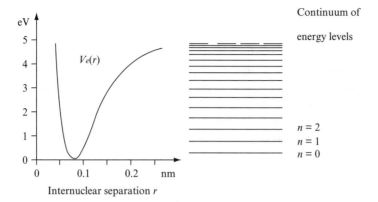

Fig. 6.4   The effective internuclear potential $V_e(r)$ and the vibrational energy levels of the hydrogen molecule. The potential energy near the minimum is approximately quadratic and acts like a harmonic oscillator potential, and the lowest vibrational levels are approximately equally spaced and given by the $E_n = (n + \frac{1}{2})\hbar\omega$. The vibrational levels become more closely spaced as the degree of excitation increases and the dissociation of the molecule gives rise to a continuum of energy levels.

If classical physics were applicable, the nuclei would have energy

$$E_{\text{classical}} = \frac{p_1^2}{2m_1} + \frac{p_2^2}{2m_2} + \frac{1}{2}kx^2.$$

where m_1 and m_2 are the masses of the nuclei and p_1 and p_2 are the magnitudes of their momenta. In the centre-of-mass frame, we can set $p_1 = p_2 = p$ and, by introducing the reduced mass

$$\mu = \frac{m_1 m_2}{m_1 + m_2},$$

we obtain

$$E_{\text{classical}} = \frac{p^2}{2\mu} + \frac{1}{2}kx^2.$$

This energy is the same as the energy of a single particle of mass μ on a spring with elastic constant k. Accordingly, we expect the vibrating nuclei in a diatomic molecule to act like a harmonic oscillator with classical frequency $\omega = \sqrt{k/\mu}$, where μ is the reduced mass of the nuclei and k is an elastic constant characterizing the strength of the molecular bond between the nuclei.

The quantum mechanical behaviour of this oscillator is described by a wave function $\Psi(x, t)$ which satisfies the Schrödinger equation

$$i\hbar\frac{\partial\Psi}{\partial t} = \left[-\frac{\hbar^2}{2\mu}\frac{\partial^2}{\partial x^2} + \frac{1}{2}k^2 x^2\right]\Psi. \qquad (6.18)$$

This equation is almost identical to Eq. (6.8), which formed the starting point for our discussion of the quantum oscillator. Indeed, if we replace the mass m by the reduced mass μ, we can apply all our results to a diatomic molecule. Most importantly, we can use Eq. (6.12) to write down an expression for the vibrational energy levels of a diatomic molecule with reduced mass μ and elastic constant k:

$$E_n = \left(n + \frac{1}{2}\right)\hbar\omega, \quad \text{where} \quad \omega = \sqrt{\frac{k}{\mu}}. \qquad (6.19)$$

The quantum number n can take on the values $0, 1, 2, \ldots$, but, when n is large, the harmonic oscillator model for molecular vibrations breaks down. This occurs when the vibrational energy becomes comparable with the dissociation energy of the molecule, as shown for the hydrogen molecule in Fig. 6.4.

A transition from one vibrational level of the molecule to another is often accompanied by the emission or absorption of electromagnetic radiation, usually in the infrared part of the spectrum. This is particularly so for diatomic molecules with two different nuclei, i.e. heteronuclear diatomic molecules. For such molecules, the electrons form an electric dipole which can strongly absorb or emit electromagnetic radiation. In fact, this mechanism leads to transitions between adjacent vibrational levels and the emission or absorption of photons with energy[4]

$$E = \hbar\sqrt{\frac{k}{\mu}}.$$

These photons give rise to a prominent spectral line with wavelength

$$\lambda = \frac{hc}{E} = 2\pi c\sqrt{\frac{\mu}{k}}. \qquad (6.20)$$

---

[4] The probability for transition from ψ_m to ψ_n induced by electric dipole radiation is proportional to $|x_{m,\,n}|^2$ where

$$x_{m,\,n} = \int_{-\infty}^{+\infty} \psi_m^*(x)x\psi_n(x)\,\mathrm{d}x.$$

By using the properties of the harmonic oscillator eigenfunctions, one can show that $x_{m,\,n} = 0$ if $|m - n| \neq 1$. (See problem 11 at the end of this chapter.)

As an example, we consider the carbon monoxide molecule. The reduced mass of the nuclei is $\mu = 6.85$ amu, and, when transitions between adjacent vibrational levels occur, infrared radiation with wavelength $\lambda = 4.6\,\mu$m is emitted or absorbed. If we substitute these values for μ and λ into Eq. (6.20), we find that the elastic constant, characterizing the strength of the bond in the carbon monoxide molecule, is $k = 1908$ Nm^{-1}.

In reality, the situation is more complex. First of all, transitions between adjacent vibrational levels have slightly different wavelengths, because the vibrational energy levels are only approximately equally spaced; as illustrated in Fig. 6.4, a harmonic oscillator potential does not exactly describe the interaction between the nuclei in a diatomic molecule. Second, the molecule may rotate and each vibrational level is really a band of closely spaced levels with different rotational energies; accordingly, there is a band of spectral lines associated with each vibrational transition.

6.5   THREE-DIMENSIONAL OSCILLATORS

We shall conclude this chapter by considering a particle of mass m in the three-dimensional harmonic oscillator potential

$$V(r) = \tfrac{1}{2}kr^2 = \tfrac{1}{2}k(x^2 + y^2 + z^2). \tag{6.21}$$

A classical particle at a distance r from the origin would experience a central force towards the origin of magnitude kr. When displaced from the origin and released, it executes simple harmonic motion with angular frequency $\omega = \sqrt{k/m}$, but more complicated motion occurs when the particle is displaced and also given a transverse velocity.

The behaviour of a quantum particle is governed by a Hamiltonian operator $\hat{H}$ which is the sum of three one-dimensional Hamiltonians:

$$\hat{H} = \hat{H}_x + \hat{H}_y + \hat{H}_z \tag{6.22}$$

where

$$\hat{H}_x = -\frac{\hbar^2}{2m}\frac{\partial^2}{\partial x^2} + \frac{1}{2}m\omega^2 x^2,$$

$$\hat{H}_y = -\frac{\hbar^2}{2m}\frac{\partial^2}{\partial y^2} + \frac{1}{2}m\omega^2 y^2,$$

$$\hat{H}_z = -\frac{\hbar^2}{2m}\frac{\partial^2}{\partial z^2} + \frac{1}{2}m\omega^2 z^2.$$

Stationary states with definite energy are represented by wave functions of the form

$$\Psi(x, y, z, t) = \psi(x, y, z)\, e^{-iEt/\hbar}, \tag{6.23}$$

where $\psi(x, y, z)$ and E satisfy the three-dimensional eigenvalue equation

$$\hat{H}\psi(x, y, z) = E\psi(x, y, z). \tag{6.24}$$

These states may be found by using the eigenvalue equations for the one-dimensional oscillators governed by $\hat{H}_x, \hat{H}_y$ and $\hat{H}_z$:

$$\hat{H}_x \psi_{n_x}(x) = \left(n_x + \tfrac{1}{2}\right)\hbar\omega\psi_{n_x}(x),$$
$$\hat{H}_y \psi_{n_y}(y) = \left(n_y + \tfrac{1}{2}\right)\hbar\omega\psi_{n_y}(y),$$
$$\hat{H}_z \psi_{n_z}(z) = \left(n_z + \tfrac{1}{2}\right)\hbar\omega\psi_{n_z}(z),$$

where the quantum numbers n_x, n_y and n_z can take on the values $0, 1, 2 \ldots \ldots$ These three equations imply that the function

$$\psi_{n_x, n_y, n_z}(x, y, z) = \psi_{n_x}(x)\psi_{n_y}(y)\psi_{n_z}(z) \tag{6.25}$$

satisfies the three-dimensional eigenvalue equation

$$\hat{H}\psi_{n_x, n_y, n_z}(x, y, z) = E_{n_x, n_y, n_z}\psi_{n_x, n_y, n_z}(x, y, z) \tag{6.26}$$

provided that

$$E_{n_x, n_y, n_z} = \left(n_x + n_y + n_z + \tfrac{3}{2}\right)\hbar\omega. \tag{6.27}$$

Thus, the eigenvalues and the eigenfunctions of the three-dimensional oscillator are labelled by three quantum numbers, n_x, n_y and n_z, each of which can take on any integer value between zero and infinity. The explicit forms of the low-lying eigenfunctions can be found by using Table 6.1. When all three quantum numbers are equal to 0, we have the ground state:

$$E_{0,0,0} = \frac{3}{2}\hbar\omega \quad \text{and} \quad \psi_{0,0,0}(x, y, z) = \left(\frac{1}{a\sqrt{\pi}}\right)^{3/2} e^{-(x^2+y^2+z^2)/2a^2},$$

where $a = \sqrt{\hbar/m\omega}$. By changing one of the quantum numbers from 0 to 1, we obtain three excited states with the same energy:

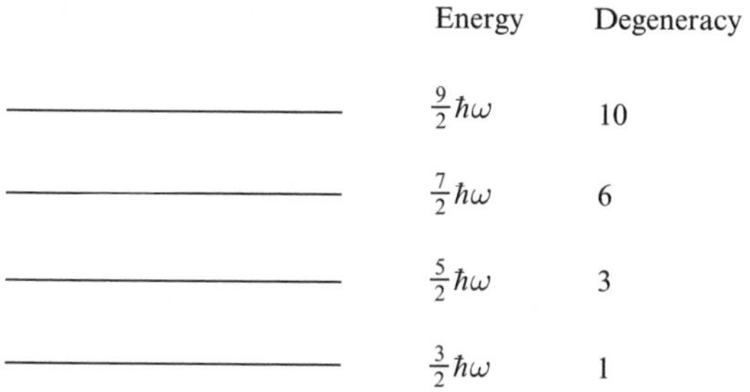

	Energy	Degeneracy
————————————	$\frac{9}{2}\hbar\omega$	10
————————————	$\frac{7}{2}\hbar\omega$	6
————————————	$\frac{5}{2}\hbar\omega$	3
————————————	$\frac{3}{2}\hbar\omega$	1

Fig. 6.5 The four lowest energy levels of a particle in a three-dimensional harmonic oscillator potential. The degeneracy of each level is denoted on the right-hand side.

$$E_{1,0,0} = \frac{5}{2}\hbar\omega \quad \text{and} \quad \psi_{1,0,0}(x,y,z) = \left(\frac{1}{a\sqrt{\pi}}\right)^{3/2} 2^{1/2}\left(\frac{x}{a}\right) e^{-(x^2+y^2+z^2)/2a^2};$$

$$E_{0,1,0} = \frac{5}{2}\hbar\omega \quad \text{and} \quad \psi_{0,1,0}(x,y,z) = \left(\frac{1}{a\sqrt{\pi}}\right)^{3/2} 2^{1/2}\left(\frac{y}{a}\right) e^{-(x^2+y^2+z^2)/2a^2};$$

$$E_{0,0,1} = \frac{5}{2}\hbar\omega \quad \text{and} \quad \psi_{0,0,1}(x,y,z) = \left(\frac{1}{a\sqrt{\pi}}\right)^{3/2} 2^{1/2}\left(\frac{z}{a}\right) e^{-(x^2+y^2+z^2)/2a^2}.$$

In a similar way we can find six states with energy $7\hbar\omega/2$, ten states with energy $9\hbar\omega/2$, and so on.

The energy levels of the three-dimensional harmonic oscillator are shown in Fig. 6.5. This diagram also indicates the degeneracy of each level, the degeneracy of an energy level being the number of independent eigenfunctions associated with the level. This degeneracy arises because the Hamiltonian for the three-dimensional oscillator has rotational and other symmetries.

6.6   THE OSCILLATOR EIGENVALUE PROBLEM

For the benefit of mathematically inclined readers we shall now discuss the problem of finding the energy eigenfunctions and eigenvalues of a one-dimensional harmonic oscillator. The method used is interesting and introduces mathematical methods which are very useful in advanced quantum mechanics. *This section may be omitted without significant loss of continuity.*

In order to simplify the task of finding the eigenvalues and eigenfunctions, we shall clean up the eigenvalue equation (6.10) and give it a gentle massage. We note that this equation contains three dimensional constants: Planck's constant $\hbar$, the classical angular frequency ω, and the mass of the confined particle m. With these constants we can construct an energy $\hbar\omega$ and a length $\sqrt{\hbar/m\omega}$. Hence, it is natural to measure the energy E in units of $\hbar\omega$

and the length x in units of $\sqrt{\hbar/m\omega}$. Accordingly, we shall rewrite Eq. (6.10) using

$$E = \epsilon\hbar\omega \quad \text{and} \quad x = q\sqrt{\frac{\hbar}{m\omega}}. \tag{6.28}$$

If we think of the eigenfunction ψ as a function of q, Eq. (6.10) becomes

$$\left[-\frac{d^2}{dq^2} + q^2\right]\psi(q) = 2\epsilon\psi(q). \tag{6.29}$$

The task of finding the eigenvalues ϵ and eigenfunctions $\psi(q)$ is made easy by noting that for any function $f(q)$

$$\left[q + \frac{d}{dq}\right]\left[q - \frac{d}{dq}\right]f(q) = q^2 f(q) - \frac{d^2 f(q)}{dq^2} - q\frac{df(q)}{dq} + \frac{d[qf(q)]}{dq}.$$

Using

$$\frac{d[qf(q)]}{dq} = f(q) + q\frac{df(q)}{dq}$$

we obtain

$$\left[q + \frac{d}{dq}\right]\left[q - \frac{d}{dq}\right]f(q) = \left[q^2 - \frac{d^2}{dq^2} + 1\right]f(q).$$

It follows that the eigenvalue equation Eq. (6.29) may be written as

$$\left[q + \frac{d}{dq}\right]\left[q - \frac{d}{dq}\right]\psi(q) = (2\epsilon + 1)\psi(q). \tag{6.30}$$

Similar considerations show that we can also write the eigenvalue equation as

$$\left[q - \frac{d}{dq}\right]\left[q + \frac{d}{dq}\right]\psi(q) = (2\epsilon - 1)\psi(q). \tag{6.31}$$

Instead of finding all the eigenfunctions at one go, as we did in Section 4.4 for the infinite square-well, we shall adopt a more elegant approach of first finding the eigenfunction of the ground state and then use this as a starting point for finding the eigenfunctions of the excited states.

The ground state

Two possible eigenvalues and eigenfunctions are immediately apparent from an inspection of the alternative expressions for the eigenvalue equation given by Eq. (6.30) and Eq. (6.31).

First, Eq. (6.30) is clearly satisfied if

$$\epsilon = -\frac{1}{2} \quad \text{and} \quad \left[q - \frac{\mathrm{d}}{\mathrm{d}q} \right] \psi(q) = 0.$$

This first-order differential equation for $\psi(q)$ has the solution

$$\psi(q) = A\,\mathrm{e}^{+q^2/2}$$

where A is a constant. But this solution must be discarded because it does not satisfy the boundary conditions, $\psi(q) \to 0$ as $q \to \pm\infty$, which are needed for a normalizable wave function.

Second, Eq. (6.31) is clearly satisfied if

$$\epsilon = +\frac{1}{2} \quad \text{and} \quad \left[q + \frac{\mathrm{d}}{\mathrm{d}q} \right] \psi(q) = 0.$$

In this case the differential equation for $\psi(q)$ has the solution

$$\psi(q) = A\,\mathrm{e}^{-q^2/2},$$

which is an acceptable eigenfunction because $\psi(q) \to 0$ as $q \to \pm\infty$. Later we shall show that this is the eigenfunction of the ground state. Accordingly, we shall use the quantum number $n = 0$ as a label and take the ground state eigenvalue and eigenfunction to be

$$\epsilon_0 = +\tfrac{1}{2} \quad \text{and} \quad u_0(q) = A_0\,\mathrm{e}^{-q^2/2}, \tag{6.32}$$

where A_0 is a normalization constant.

If we use Eq. (6.28) to express the dimensionless variables ϵ and q in terms of the dimensional variables E and x, we find that the ground state of a harmonic oscillator with angular frequency ω has energy

$$E_0 = \tfrac{1}{2}\hbar\omega \tag{6.33}$$

and that its eigenfuction, as a function of x, is given by

$$\psi_0(x) = N_0\, e^{-x^2/2a^2}, \quad \text{where} \quad a = \sqrt{\frac{\hbar}{m\omega}}. \tag{6.34}$$

The constant N_0 is a normalization constant.

Excited states

We shall now find the solutions which describe the excited states of the oscillator. The first step is to focus on the nth excited state. Its eigenfunction and eigenvalue satisfy an eigenvalue equation which we may write down using either Eq. (6.30) or Eq. (6.31). We shall choose Eq. (6.30) and write

$$\left[q + \frac{d}{dq}\right]\left[q - \frac{d}{dq}\right]\psi_n(q) = (2\epsilon_n + 1)\psi_n(q).$$

We now operate with $[q - \frac{d}{dq}]$ on both sides to give

$$\left[q - \frac{d}{dq}\right]\left[q + \frac{d}{dq}\right]\left(\left[q - \frac{d}{dq}\right]\psi_n(q)\right) = (2\epsilon_n + 1)\left(\left[q - \frac{d}{dq}\right]\psi_n(q)\right)$$

and then compare with the form of eigenvalue equation given by Eq. (6.31); i.e. we compare with

$$\left[q - \frac{d}{dq}\right]\left[q + \frac{d}{dq}\right]\psi(q) = (2\epsilon - 1)\psi(q).$$

This comparison shows that the function $\left[q - \frac{d}{dq}\right]\psi_n(q)$ is an eigenfunction $\psi(q)$ with an eigenvalue ϵ given by $(2\epsilon - 1) = (2\epsilon_n + 1)$; i.e. it is an eigenfunction with eigenvalue $\epsilon = \epsilon_n + 1$.

We have discovered that the operator $[q - \frac{d}{dq}]$ has the role of an energy raising operator. When it operates on the eigenfunction $\psi_n(q)$ with eigenvalue ϵ_n, it gives the eigenfunction $\psi_{n+1}(q)$ with eigenvalue $\epsilon_{n+1} = \epsilon_n + 1$. It follows that the ground-state eigenvalue and eigenfunction given by Eq. (6.32) may be used as the starting point for generating an infinite set of eigenvalues and eigenfunctions which describe excited states of the harmonic oscillator. The first excited state is described by

$$\epsilon_1 = \frac{3}{2} \quad \text{and} \quad \psi_1(q) = A_1\left[q - \frac{d}{dq}\right]e^{-q^2/2}, \tag{6.35}$$

and the second excited state is described by

$$\epsilon_1 = \frac{5}{2} \quad \text{and} \quad \psi_2(q) = A_2 \left[q - \frac{\mathrm{d}}{\mathrm{d}q} \right]^2 \mathrm{e}^{-q^2/2}, \tag{6.36}$$

and so on, *ad infinitum*. Thus, we can construct an infinite sequence of energy eigenvalues and eigenfunctions; they may be labelled by the quantum numbers $n = 0, 1, 2, \ldots$ and they are given by

$$\epsilon_n = n + \frac{1}{2} \quad \text{and} \quad \psi_n(q) = A_n \left[q - \frac{\mathrm{d}}{\mathrm{d}q} \right]^n \mathrm{e}^{-q^2/2}. \tag{6.37}$$

By using Eq. (6.28) to express ϵ in terms of E, we find that the energy of the nth level of a harmonic oscillator with angular frequency ω is

$$E_n = \left(n + \tfrac{1}{2} \right) \hbar \omega. \tag{6.38}$$

And by expressing q in terms of x we find, after a little algebra, that the nth eigenfunction has the form

$$\psi_n(x) = N_n H_n \left(\frac{x}{a} \right) \mathrm{e}^{-x^2/2a^2}, \quad \text{where} \quad a = \sqrt{\frac{\hbar}{m\omega}}. \tag{6.39}$$

The constant N_n is a normalization constant, and the function H_n, a polynomial of order n in x/a, is called a Hermite polynomial. The normalized eigenfunctions for the $n = 0, 1, 2, 3$ and 4 states are listed in Table 6.1.

Is E_0 really the lowest energy?

We have one item of unfinished mathematics. We have yet to show that $E_0 = \epsilon_0 \hbar \omega = \frac{1}{2} \hbar \omega$ is really the lowest energy of a harmonic oscillator with angular frequency ω.

In problem 1 at the end of the chapter, we shall show that the Heisenberg uncertainty principle implies that the energy of the oscillator cannot be less that $E_0 = \frac{1}{2} \hbar \omega$. In addition, in problem 9, we shall show that there is an energy lowering operator for the harmonic oscillator, but that this operator cannot be used to construct an eigenfuction with energy less than $E_0 = \frac{1}{2} \hbar \omega$. This operator has the form $[q + \frac{\mathrm{d}}{\mathrm{d}q}]$. When it operates on the eigenfunction $\psi_n(q)$ with eigenvalue ϵ_n, it yields the eigenfunction $\psi_{n-1}(q)$ with eigenvalue ϵ_{n-1}. But when this operator acts on the eigenfunction $\psi_0(q)$ it gives zero, i.e.

$$\left[q + \frac{\mathrm{d}}{\mathrm{d}q} \right] \psi_0(q) = 0. \tag{6.40}$$

Mathematical properties of the oscillator eigenfunctions

The eigenfunctions $\psi_n(x)$, like the eigenfunctions of any Hamiltonian, form a complete orthonormal set of basis functions. As described in Section 4.6, they satisfy the condition

$$\int_{-\infty}^{+\infty} \psi_m^*(x)\psi_n(x)\ \mathrm{d}x = \delta_{m,n}, \tag{6.41}$$

where $\delta_{m,n} = 1$ if $m = n$ and $\delta_{m,n} = 0$ if $m \neq n$. Moreover, they can be used as basis functions for a generalized Fourier series: any function $f(x)$ may be represented by the series

$$f(x) = \sum_{n=0,1,\dots} c_n\psi_n(x), \quad \text{where} \quad c_n = \int_{-\infty}^{+\infty} \psi_n^*(x)f(x)\ \mathrm{d}x. \tag{6.42}$$

This means that the general solution of the time-dependent Schrödinger equation for the harmonic oscillator, Eq. (6.8), has the form

$$\Psi(x,t) = \sum_{n=0,1,2\dots} c_n\psi_n(x)\,e^{-iE_n t/\hbar}, \tag{6.43}$$

where c_n are complex constants.

PROBLEMS 6

1. In this question the Heisenberg uncertainty principle

$$\Delta x\ \Delta p \geq \tfrac{1}{2}\hbar$$

is used to derive a lower bound for the energy of a particle of mass m in a harmonic oscillator potential with classical angular frequency ω.

(a) Note that the expectation value of the energy of the particle is given by

$$\langle E \rangle = \frac{\langle p^2 \rangle}{2m} + \frac{1}{2}m\omega^2\langle x^2 \rangle$$

and show that, if the average position and momentum of the particle are both zero, its energy has an expectation value which satisfies the inequality

$$\langle E \rangle \geq \frac{\hbar^2}{8m(\Delta x)^2} + \frac{1}{2}m\omega^2(\Delta x)^2.$$

(b) Show that the minimum value of a function of the form

$$F(\Delta x) = \frac{A^2}{(\Delta x)^2} + B^2(\Delta x)^2$$

is $2AB$.

(c) Hence show that the expectation value of the energy of a particle in a harmonic oscillator well satisfies the inequality

$$\langle E \rangle \geq \tfrac{1}{2}\hbar\omega.$$

2. By reference to the properties of the Gaussian distribution given in problem 2 at the end of Chapter 2, show that the position probability density for a particle of mass m in the ground state of a harmonic oscillator with angular frequency ω is a Gaussian probability distribution with standard deviation $\sigma = \sqrt{\hbar/2m\omega}$.

3. Find the amplitude of oscillation A for a classical particle with the same energy as a quantum particle in the ground state of the harmonic oscillator. Write down an expression for the probability of finding the quantum particle in the classically forbidden region $|x| > A$.

4. Consider the potential

$$V(x) = \begin{cases} \infty & \text{if } x < 0 \\ \tfrac{1}{2}m\omega^2 x^2 & \text{if } x > 0, \end{cases}$$

which describes an elastic spring which can be extended but not compressed.

By referring to the eigenfunctions of the harmonic oscillator potential shown in Fig. 6.2, sketch the eigenfunctions of the ground and first excited states of this new potential.

What are the energies of the ground and first excited states?

5. At time $t = 0$ a particle in a harmonic oscillator potential $V(x) = \tfrac{1}{2}m\omega^2 x^2$ has a wave function

$$\Psi(x, 0) = \frac{1}{\sqrt{2}}[\psi_0(x) + \psi_1(x)]$$

where $\psi_0(x)$ and $\psi_1(x)$ are real, normalized and orthogonal eigenfunctions for the ground and first-excited states of the oscillator.

(a) Write down an expression for $\Psi(x, t)$, the wave function at time t.

(b) Show that $\Psi(x, t)$ is a normalized wave function.

(c) Use your knowledge of the energy levels of the harmonic oscillator potential to show that the probability density $|\Psi(x, t)|^2$ oscillates with angular frequency ω.

(d) Show that the expectation value of x has the form

$$\langle x \rangle = A \cos \omega t, \quad \text{where} \quad A = \int_{-\infty}^{+\infty} \psi_0(x)\, x\, \psi_1(x)\, dx.$$

6. Consider the harmonic oscillator quantum state with the wave function

$$\Psi(x, t) = \sqrt{\tfrac{1}{3}}[\psi_0(x)\,e^{-iE_0 t/\hbar} + \psi_1(x)\,e^{-iE_1 t/\hbar} + \psi_2(x)\,e^{-iE_2 t/\hbar}]$$

where $\psi_0(x), \psi_1(x)$ and $\psi_2(x)$ are taken as real, normalized eigenfunctions of the harmonic oscillator with energy E_0, E_1 and E_2, respectively.

(a) What is the expectation value of the energy?

(b) What is the uncertainty in the energy?

(c) Show that the probability for the position of the particle has the form

$$|\Psi(x, t)|^2 = A(x) + B(x) \cos \omega t + C(x) \cos 2\omega t,$$

and find the functions $A(x)$, $B(x)$ and $C(x)$.

7. The transitions between adjacent vibrational levels of the NO molecule give rise to infrared radiation with wavelength $\lambda = 5.33\ \mu$m. Find the elastic constant k characterizing the strength of the bond between the nuclei in the NO molecule. (The reduced mass of the NO molecule is 7.46 amu.)

8. In answering this question, you may assume that $\psi_n(x)$ is the energy eigen-function of a particle of mass m in a one-dimensional harmonic oscillator potential $V(x) = \tfrac{1}{2}m\omega^2 x^2$ with energy $E_n = (n + \tfrac{1}{2})\hbar\omega$.

Consider a particle of mass m in a two-dimensional harmonic oscillator potential with an energy eigenvalue equation of the form:

$$\left[-\frac{\hbar^2}{2m}\left(\frac{\partial^2}{\partial x^2} + \frac{\partial^2}{\partial y^2} \right) + \frac{1}{2}m\omega^2(x^2 + y^2) \right] \psi_{n_x, n_y}(x, y) = E_{n_x, n_y} \psi_{n_x, n_y}(x, y).$$

(a) Verify that

$$\psi_{n_x, n_y}(x, y) = \psi_{n_x}(x)\psi_{n_y}(y),$$

with $n_x = 0, 1, 2, \ldots$ and $n_y = 0, 1, 2, \ldots$, is an energy eigenfunction
with energy $E_{n_x, n_y} = (n_x + n_y + 1)\hbar\omega$.

(b) Draw an energy level diagram and indicate the degeneracy of the
energy levels.

(c) By expressing $\psi_{1,0}$ and $\psi_{0,1}$ in plane polar coordinates (r, ϕ), find
functions $\psi_a(r, \phi)$ and $\psi_b(r, \phi)$ which obey the equations

$$-i\hbar\frac{\partial \psi_a}{\partial \phi} = +\hbar\psi_a \quad \text{and} \quad -i\hbar\frac{\partial \psi_b}{\partial \phi} = -\hbar\psi_b.$$

The remaining questions are for readers who studied Section 6.6.

9. In this question you are asked to show that when the operator $\left[q + \frac{d}{dq} \right]$ acts
   on the eigenfunction $\psi_n(q)$ with eigenvalue ϵ_n, it gives the eigenfunction
   $\psi_{n-1}(q)$ with eigenvalue $\epsilon_{n-1} = \epsilon_n - 1$.
   Consider the nth state of the oscillator with eigenvalue ϵ_n and eigenfunction ψ_n. Using the form of the eigenvalue equation given by Eq. (6.31),
   write

$$\left[q - \frac{d}{dq} \right]\left[q + \frac{d}{dq} \right]\psi_n(q) = (2\epsilon_n - 1)\psi_n(q).$$

Now operate with $\left[q + \frac{d}{dq} \right]$ on both sides, compare with the form of the
eigenvalue equation given by Eq. (6.30) and show that the function
$\left[q + \frac{d}{dq} \right]\psi_n(q)$ is an eigenfunction $\psi(q)$ with an eigenvalue ϵ given by
$\epsilon = \epsilon_n - 1$.

10. In general, the eigenfunctions $\psi_n(q)$ and $\psi_{n+1}(q)$ are related by

$$\left[q - \frac{d}{dq} \right]\psi_n(q) = a_n\psi_{n+1}(q)$$

where a_n is a constant. Show that $\psi_n(q)$ and $\psi_{n+1}(q)$ have the same
normalization if

$$|a_n|^2 = 2(n+1).$$

[Hint: Write down the normalization integrals, note that integration by parts can be used to show that

$$\int_{-\infty}^{+\infty} \left[\frac{df(q)}{dq}\right] g(q)\, dq = -\int_{-\infty}^{+\infty} f(q)\left[\frac{dg(q)}{dq}\right] dq$$

if the function $f(q)g(q)$ goes to zero as $q \to \pm\infty$, and use Eq. (6.30).]

Similarly, show that the eigenfunctions $\psi_n(q)$ and $\psi_{n-1}(q)$, that are related by

$$\left[q + \frac{d}{dq}\right]\psi_n(q) = b_n\psi_{n-1}(q),$$

have the same normalization if

$$|b_n|^2 = 2n.$$

11. Consider eigenfunctions ψ_n of the harmonic oscillator which obey the normalization condition

$$\int_{-\infty}^{+\infty} |\psi_n(q)|^2\, dq = 1.$$

(a) Show that

$$\int_{-\infty}^{+\infty} \psi_m^*(q)\, q\, \psi_n(q)\, dq = \tfrac{1}{2}(a_n\delta_{m,\,n+1} + b_n\delta_{m,\,n-1})$$

where a_n and b_n are defined in problem 10.

(b) Show that

$$\int_{-\infty}^{+\infty} \psi_n^*(q)\, q^2\, \psi_n(q)\, dq = \left(n + \tfrac{1}{2}\right)$$

and that

$$\int_{-\infty}^{+\infty} \psi_n^*(q)\left[-\frac{d^2}{dq^2}\right]\psi_n(q)\, dq = \left(n + \tfrac{1}{2}\right).$$

(c) Now consider the eigenfunction ψ_n as a function of x and normalized so that

$$\int_{-\infty}^{+\infty} |\psi_n(x)|^2 \, dx = 1.$$

Verify that

$$\langle x \rangle = 0 \quad \text{and} \quad \langle x^2 \rangle = \left(n + \tfrac{1}{2}\right) a^2$$

and that

$$\langle p \rangle = 0 \quad \text{and} \quad \langle p^2 \rangle = \left(n + \tfrac{1}{2}\right) \hbar/a^2.$$

[Hint: Write

$$q = \frac{1}{2}\left[q + \frac{d}{dq}\right] + \frac{1}{2}\left[q - \frac{d}{dq}\right] \quad \text{and} \quad \frac{d}{dq} = \frac{1}{2}\left[q + \frac{d}{dq}\right] - \frac{1}{2}\left[q - \frac{d}{dq}\right],$$

use the fact that $[q - \frac{d}{dq}]$ and $[q + \frac{d}{dq}]$ are energy raising and lowering operators, note that the eigenfunctions obey the orthogonality relation Eq. (6.41), and use Eq. (6.30) and Eq. (6.31).]

7

Observables and operators

Operators have to be used in quantum mechanics to describe observable quantities because measurements may have uncertain outcomes. In Chapter 3 we used the operators

$$\hat{\mathbf{r}} = \mathbf{r} \quad \text{and} \quad \hat{\mathbf{p}} = -i\hbar\nabla$$

to calculate the expectation values and the uncertainties in the position and momentum of a particle. In Chapter 4 we used the Hamiltonian operator

$$\hat{H} = -\frac{\hbar^2}{2m}\nabla^2 + V(\mathbf{r})$$

to explore the energy properties of a particle. And in the next chapter we shall consider in detail a fourth operator, the operator describing the orbital angular momentum of a particle,

$$\hat{\mathbf{L}} = \hat{\mathbf{r}} \times \hat{\mathbf{p}}.$$

In this chapter we shall consider some physical properties of observables in quantum mechanics and link these properties to the mathematical properties of the operators which describe observables. In so doing, concepts that were implicit in the use of operators in earlier chapters will be clarified and developed. This chapter will deal with concepts that are more abstract and mathematical than those encountered elsewhere in this book. *It may be omitted without significant loss of continuity.*

7.1   ESSENTIAL PROPERTIES

In Chapter 4 we developed a mathematical description of energy measurement based upon the properties of the Hamiltonian operator $\hat{H}$. We explained why an eigenfunction of $\hat{H}$ belonging to an eigenvalue E represents a quantum state with sharply defined energy E. We also explained why any wave function Ψ can be expressed as a linear superposition of energy eigenfunctions. In particular, when the operator $\hat{H}$ only has eigenfunctions ψ_n with discrete eigenvalues E_n, any wave function may be written as

$$\Psi(\mathbf{r}, t) = \sum_n c_n \psi_n(\mathbf{r}) \, e^{-iE_n t/\hbar},$$

where the coefficients c_n are probability amplitudes, such that, if the energy is measured, $|c_n|^2$ is the probability of an outcome E_n. When the Hamiltonian gives rise to a continuum of energy eigenvalues, the general wave function involves an integral over the continuous energy variable which labels the eigenfunctions.

We shall use the Hamiltonian operator as a prototype for all operators which describe observables in quantum mechanics. We shall consider a general observable A described by an operator $\hat{A}$, take note of the fundamental concepts discussed in Chapter 4 and highlight the essential mathematical properties of the operator $\hat{A}$. They are the following:

- The operator $\hat{A}$ must be a *linear operator*. This means that, if the action of $\hat{A}$ on the wave functions Ψ_1 and Ψ_2 is given by

$$\hat{A}\Psi_1 = \Phi_1 \quad \text{and} \quad \hat{A}\Psi_2 = \Phi_2,$$

then the action of $\hat{A}$ on the wave function $c_1\Psi_1 + c_2\Psi_2$, where c_1 and c_2 are two arbitrary complex numbers, is given by

$$\hat{A}(c_1\Psi_1 + c_2\Psi_2) = c_1\Phi_1 + c_2\Phi_2. \tag{7.1}$$

This abstract property is satisfied by the operators for $\hat{H}$, $\hat{\mathbf{r}}$, $\hat{\mathbf{p}}$ and by all other operators which describe observables in quantum mechanics. It ensures that these observables are consistent with the principle of linear superposition which asserts that any quantum state is a linear superposition of other quantum states.

- The operator $\hat{A}$ must be a *Hermitian operator* which means that it obeys the condition

$$\int \Psi_1^* \hat{A} \; \Psi_2 \; d^3\mathbf{r} = \int (\hat{A}\Psi_1)^* \; \Psi_2 \; d^3\mathbf{r}, \qquad (7.2)$$

where Ψ_1 and Ψ_2 are any two wave functions; the brackets in the term $(\hat{A}\Psi_1)^*$ mean that the operator $\hat{A}$ only acts on the wave function Ψ_1 and the complex conjugate of the result is taken. This mathematical property ensures that the expectation value of the observable,

$$\langle A \rangle = \int \Psi^* \; \hat{A} \; \Psi \; d^3\mathbf{r},$$

is real for any wave function Ψ.

It also ensures that the eigenvalues of the operator $\hat{A}$ are real. There may be eigenfunctions $\psi_{a_n}(\mathbf{r})$ with discrete eigenvalues a_n given by

$$\hat{A}\psi_{a_n} = a_n \psi_{a_n}$$

and/or eigenfunctions $\psi_{a'}$ with continuous eigenvalues a' given by

$$\hat{A}\psi_{a'} = a' \psi_{a'}.$$

In Chapter 4 we argued that the possible outcomes of an energy measurement are energy eigenvalues. Identical arguments imply that the possible outcomes of a measurement of the observable A are the eigenvalues of the operator $\hat{A}$. Because the outcomes of all measurements are real numbers, the eigenvalues of the operator $\hat{A}$ must be real numbers.

- Finally, the operator $\hat{A}$ must describe an observable which is always measurable. Specifically, we must be able to predict the outcomes of a measurement of A and the probability of each of these outcomes. This is only possible if the eigenfunctions of the operator $\hat{A}$ form a complete set of basis functions so that any wave function $\Psi(\mathbf{r}, t)$ can be written as

$$\Psi(\mathbf{r}, t) = \sum_n c_{a_n}(t)\psi_{a_n}(\mathbf{r}) + \int c(a', t)\psi_{a'}(\mathbf{r}) \, da'. \qquad (7.3)$$

In this expression $c_{a_n}(t)$ and $c(a', t)$ are probability amplitudes for the observable A. If a measurement takes place at time t on a particle with wave function Ψ, then $|c_{a_n}(t)|^2$ is the probability of outcome a_n and $|c(a', t)|^2 da'$ is the probability of an outcome between a' and $a' + da'$.[1]

---

[1] To keep the presentation as simple as possible, we have ignored the complications that arise when there is degeneracy, i.e. more than one eigenfunction with the same eigenvalue, and we have not addressed, at this stage, the normalization and orthogonality of eigenfunctions with continuous eigenvalues.

7.2   POSITION AND MOMENTUM

In Chapter 3 we introduced the operators for the position and momentum of a particle and described how they can be used to calculate expectation values and uncertainties, but we have not yet explicitly considered the eigenvalues and eigenfunctions of these operators. We shall do so by considering a particle moving in one dimension. We shall denote an eigenfunction for a particle with position eigenvalue x' by $\psi_{x'}(x)$ and an eigenfunction for a particle with momentum eigenvalue p' by $\psi_{p'}(x)$.

Eigenfunctions for position

The position eigenfunction satisfies the eigenvalue equation

$$\hat{x}\psi_{x'}(x) = x'\psi_{x'}(x) \tag{7.4}$$

which may be rewritten using $\hat{x} = x$ to give

$$x\psi_{x'}(x) = x'\psi_{x'}(x). \tag{7.5}$$

This equation states that $\psi_{x'}(x)$ multiplied by x is the same as $\psi_{x'}(x)$ multiplied by the eigenvalue x'. This is only possible if $\psi_{x'}(x)$ is a very peculiar function which is infinitely peaked at $x = x'$. Such a function is normally written as

$$\psi_{x'}(x) = \delta(x - x'), \tag{7.6}$$

where $\delta(x - x')$ is a Dirac delta function.

A Dirac delta function can be considered as the limiting case of more familiar functions. For example, the function

$$\delta_\epsilon(x - x') = \begin{cases} 1/2\epsilon & \text{if } |x - x'| < \epsilon \\ 0 & \text{if } |x - x'| > \epsilon, \end{cases}$$

which is illustrated in Fig. 7.1, can behave like a Dirac delta function because it becomes increasingly high and narrow at $x = x'$ as ϵ tends to zero. In fact, the defining property of a Dirac delta function is that, for any function $f(x)$,

$$\int_{-\infty}^{+\infty} f(x)\delta(x - x') \, dx = f(x'). \tag{7.7}$$

This definition is satisfied by $\delta_\epsilon(x - x')$ with $\epsilon \to 0$ because, when this limit is taken, the function becomes increasingly high and narrow at $x = x'$ and the area under the function remains equal to unity.

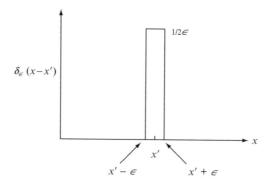

Fig. 7.1 A function $\delta_\epsilon(x - x')$ which behaves like a Dirac delta function $\delta(x - x')$. As the parameter ϵ tends to zero, the area under the function $\delta_\epsilon(x - x')$ remains equal to unity and the function becomes increasingly high and narrow at $x = x'$.

The eigenfunctions of position, like those of any observable, form a complete set of basis functions. In particular, any wave function $\Psi(x, t)$ may be written as

$$\Psi(x, t) = \int_{-\infty}^{+\infty} c(x', t)\psi_{x'}(x) \, dx' \tag{7.8}$$

where the function $c(x', t)$ is a position probability amplitude; i.e. $|c(x', t)|^2 dx'$ is the probability of finding the particle between x' and $x' + dx'$ at time t. If we use Eq. (7.6) and the definition of the Dirac delta function, Eq. (7.7), we find

$$\Psi(x, t) = \int_{-\infty}^{+\infty} c(x', t)\delta(x' - x) \, dx' = c(x, t).$$

This equation confirms the assumption we made in Chapter 3 that a wave function $\Psi(x, t)$ is a probability amplitude for the position of the particle.

Eigenfunctions for momentum

An eigenfunction for a particle with momentum eigenvalue p', $\psi_{p'}(x)$, satisfies the equation

$$\hat{p}\psi_{p'}(x) = p'\psi_{p'}(x). \tag{7.9}$$

This may be rewritten, by using

$$\hat{p} = -i\hbar\frac{\partial}{\partial x},$$

to give the differential equation

$$-i\hbar \frac{\partial \psi_{p'}(x)}{\partial x} = p' \psi_{p'}(x),$$ (7.10)

which has solutions of the form

$$\psi_{p'}(x) = \frac{1}{\sqrt{2\pi\hbar}}\, e^{ip'x/\hbar}.$$ (7.11)

As expected, an eigenfunction with momentum p' is a plane wave with wave number $k' = p'/\hbar$ and wavelength $\lambda' = h/p'$. The constant $1/\sqrt{2\pi\hbar}$ is a useful convention which ensures that the momentum and position eigenfunctions obey similar normalization conditions.

Because the momentum eigenfunctions form a complete set of basis functions, any wave function $\Psi(x,t)$ may be written as

$$\Psi(x,t) = \int_{-\infty}^{+\infty} c(p',t)\psi_{p'}(x)\, dp',$$ (7.12)

where the function $c(p',t)$ is a momentum probability amplitude; i.e. $|c(p',t)|^2 dp'$ is the probability that the measured momentum of the particle at time t is between p' and $p' + dp'$. If we use Eq. (7.11), we obtain

$$\Psi(x,t) = \frac{1}{\sqrt{2\pi\hbar}} \int_{-\infty}^{+\infty} c(p',t)\, e^{ip'x/\hbar}\, dp',$$

which, apart from differences in notation, is identical to the equation

$$\Psi(x,t) = \frac{1}{\sqrt{2\pi\hbar}} \int_{-\infty}^{+\infty} \tilde{\Psi}(p,t)\, e^{+ipx/\hbar}\, dp,$$

which was introduced in Chapter 3 when we made the assumption that the Fourier transform of the wave function $\tilde{\Psi}(p,t)$ is the probability amplitude for momentum; see Eq. (3.19) in Section 3.3. We now see that the assumption we made in Chapter 3 is consistent with the general description of observables being developed in this chapter.

Delta function normalization

For the benefit of more mathematically inclined readers we shall briefly discuss a minor mathematical problem with the normalization of the eigenfunctions for position and momentum. This problem arises because both these eigenfunctions have continuous eigenvalues; a similar problem afflicts energy eigenfunctions when energy eigenvalues are continuous.

If we use Eq. (7.6) and the definition of the Dirac delta function, Eq. (7.7), we find that the position eigenfunctions obey the condition

$$\int_{-\infty}^{+\infty} \psi_{x'}^*(x)\psi_{x''}(x) \, \mathrm{d}x = \delta(x' - x''). \tag{7.13}$$

Because the delta function is zero when $x' \neq x''$ and infinite when $x' = x''$, we conclude that position eigenfunctions are mutually orthogonal but that they cannot be normalized to unity.

Similarly, Fourier transform techniques may be used to show that the momentum eigenfunctions given by Eq. (7.11) obey the condition

$$\int_{-\infty}^{+\infty} \psi_{p'}^*(x)\psi_{p''}(x) \, \mathrm{d}x = \delta(p' - p''). \tag{7.14}$$

Hence momentum eigenfunctions are also mutually orthogonal but not normalizable to unity.

We recall that a wave function must be normalized to unity, as in Eq. (3.17), if it describes a particle which can always be found somewhere. Thus, strictly speaking, the position and momentum eigenfunctions, given by Eq. (7.6) and Eq. (7.11), cannot be used to describe a physically acceptable wave function. However, normalizable wave functions can be formed by taking linear superpositions of these eigenfunctions and these wave functions are wave packets which can be used to describe particles with very small uncertainties in position or in momentum.

Finally, we note that the use of delta functions can be avoided by imagining the particle to be in a box of large dimensions. When this approach is adopted, normalized eigenfunctions of position and momentum can be constructed.

7.3   COMPATIBLE OBSERVABLES

In classical physics it is always possible, in principle, to have precise knowledge of all the observable properties of a system at a given instant of time. For example, the specific values of the position and momentum of a particle can be used to find the value of any other dynamical observable. We now know that the comprehensive precision of classical physics cannot be realized in practice because measurement is an activity which may affect the system. Accordingly, quantum states of motion may only be specified by data which, in classical physics, would be deemed limited or imprecise. If we want precision in quantum physics we have to select a subset of observables which can be determined without mutual interference or contradiction. Such observables are called compatible observables. When a complete set of compatible observables is specified, we can in principle write down a wave function that completely describes the quantum state. All the information on the complete set of observables and all the information on the probabilities for the uncertain values of other observables will be contained in this wave function.

To illustrate these general ideas we shall first consider a particle moving in one dimension and then a particle moving in three dimensions.

A classical state for a particle in one dimension may be defined by specifying two observables, the position and the momentum. In contrast, a quantum state for a particle in one dimension is completely defined if one observable is precisely specified. For example, a quantum state can be defined by a precise position or by a precise momentum. It could also be defined by precise energy; such a state is particularly useful because it is a stationary state, a state with no time-dependent observable properties.

More observables are needed to specify a state of a particle moving in three dimensions. A classical state is defined by six observables, the three position coordinates and the three components of momentum, but a quantum state can be defined by only three precisely specified observables. We could, for example, specify the x, y and z coordinates and have three uncertain components of momentum, or we could, for example, specify the x and y coordinates and the z component of the momentum and live with uncertain momentum in the x and y directions and an uncertain z coordinate. However, because states with definite energy are stationary states, it is often most useful to specify the energy and two other observables.

7.4   COMMUTATORS

The role of compatible and non-compatible observables in quantum mechanics can be made clearer by introducing the mathematical concept of *a commutator of two operators*.

The commutator of two operators $\hat{A}$ and $\hat{B}$ is defined by

$$[\hat{A}, \hat{B}] \equiv \hat{A}\hat{B} - \hat{B}\hat{A}. \qquad (7.15)$$

It is a useful concept in the mathematics of operators because, as we shall show, the order in which two operators act upon a function is important. It is a useful concept in quantum physics because its value can be used to determine whether observables are compatible or non-compatible. We shall show that two observables A and B, described by the operators $\hat{A}$ and $\hat{B}$, are non-compatible if

$$[\hat{A}, \hat{B}] \neq 0$$

and that they are compatible if

$$[\hat{A}, \hat{B}] = 0.$$

This general statement is best understood by reconsidering the quantum states for a particle in one dimension and in three dimensions.

A particle in one dimension

We can evaluate the commutator of the position and momentum operators for a particle in one dimension by considering

$$(\hat{x}\hat{p} - \hat{p}\hat{x})\Psi(x, t),$$

where $\Psi(x, t)$ is any wave function of the particle. This is non-zero because the order of $\hat{x}$ and $\hat{p}$ matters. Specifically, we have

$$\hat{x}\hat{p}\ \Psi(x, t) = x\left(-i\hbar\frac{\partial}{\partial x}\right)\Psi(x, t)$$

and

$$\hat{p}\hat{x}\ \Psi(x, t) = \left(-i\hbar\frac{\partial}{\partial x}\right)x\ \Psi(x, t) = -i\hbar\Psi(x, t) + x\left(-i\hbar\frac{\partial}{\partial x}\right)\Psi(x, t),$$

so that

$$(\hat{x}\hat{p} - \hat{p}\hat{x})\Psi(x, t) = i\hbar\Psi(x, t). \tag{7.16}$$

Because this is true for any wave function $\Psi(x, t)$, we conclude the operation defined by $(\hat{x}\hat{p} - \hat{p}\hat{x})$ is always a multiplication by the number $i\hbar$; in brief, we conclude that the commutator of $\hat{x}$ and $\hat{p}$ is

$$[\hat{x}, \hat{p}] = i\hbar. \tag{7.17}$$

This relation is so important in quantum mechanics that it is called the *canonical commutation relation*.

We can reveal the physical significance of the canonical commutation relation by assuming the impossible: the existence of a simultaneous eigenfunction of position and momentum $\psi_{x'p'}(x)$ satisfying the eigenvalue equations

$$\hat{x}\psi_{x'p'}(x) = x'\psi_{x'p'}(x) \quad \text{and} \quad \hat{p}\psi_{x'p'}(x) = p'\psi_{x'p'}(x). \tag{7.18}$$

Such an eigenfunction, if it existed, would represent a quantum state with sharply defined position and momentum, x' and p'.

Let us consider the action of the commutator of $\hat{x}$ and $\hat{p}$ on this hypothetical simultaneous eigenfunction. If we use the eigenvalue equations (7.18) we obtain

$$[\hat{x}, \hat{p}]\psi_{x'p'}(x) = (\hat{x}\hat{p} - \hat{p}\hat{x})\psi_{x'p'}(x) = (x'p' - p'x')\psi_{x'p'}(x) = 0.$$

If we use the canonical commutation relation, Eq. (7.17), we obtain

$$[\hat{x}, \hat{p}]\psi_{x'p'}(x) = i\hbar\psi_{x'p'}(x).$$

These two results imply that

$$i\hbar\psi_{x'p'}(x) = 0.$$

Thus, we have assumed the existence of a simultaneous eigenfunction of position and momentum $\psi_{x'p'}(x)$ and shown that it must be zero for all x. In other words, we have shown that a quantum state with definite position and momentum cannot exist. We emphasize that the mathematical reason for the non-existence of such a state, and hence the non-compatibility of position and momentum, is that the commutator of $\hat{x}$ and $\hat{p}$ is non-zero.

Moreover, the degree of non-compatibility of position and momentum, as expressed by the Heisenberg uncertainty principle Eq. (1.15), may be derived using the canonical commutation relation (7.17). To fully understand this derivation, readers need to know about the properties of *Hermitian operators* and an inequality called the *Schwarz inequality*, and they can gain the necessary understanding by working through problems 5 and 6 at the end of this chapter. The key steps in the derivation of the Heisenberg uncertainty principle are as follows:

The square of the uncertainties, or variances, in the position and momentum of a particle with normalized wave function $\Psi(x, t)$ are given by

$$(\Delta x)^2 = \int_{-\infty}^{+\infty} \Psi^* \, (\widehat{\Delta x})^2 \, \Psi \, dx \quad \text{and} \quad (\Delta p)^2 = \int_{-\infty}^{+\infty} \Psi^* \, (\widehat{\Delta p})^2 \, \Psi \, dx,$$

where the operators $\widehat{\Delta x}$ and $\widehat{\Delta p}$ are defined as

$$\widehat{\Delta x} \equiv \hat{x} - \langle x \rangle \quad \text{and} \quad \widehat{\Delta p} \equiv \hat{p} - \langle p \rangle.$$

By using the canonical commutation relation (7.17), we can easily show that these operators obey the commutation relation

$$[\widehat{\Delta x}, \widehat{\Delta p}] = i\hbar. \tag{7.19}$$

We can now apply the general results derived in problem 4 and 5 to give

$$(\Delta x)^2(\Delta p)^2 \geq \frac{1}{4}\left|\int_{-\infty}^{+\infty} \Psi^* \, [\widehat{\Delta x}, \widehat{\Delta p}] \, \Psi \, dx\right|^2,$$

which can be simplified, using the commutation relation (7.19) and the normalization condition for the wave function Ψ, to give

$$(\Delta x)^2(\Delta p)^2 \geq \frac{\hbar^2}{4}$$

or

$$\Delta x \, \Delta p \geq \frac{\hbar}{2}.$$

Thus, the Heisenberg uncertainty principle, which was introduced in Chapter 1 to illustrate the inherent uncertainties associated with position and momentum measurements, can be derived by assuming that position and momentum observables are described by operators that obey the canonical commutation relation (7.17).

A particle in three dimensions

By considering a particle moving in three dimensions, we can illustrate the connection between commutators and compatible observables. In this case, a unique quantum state is defined by specifying three compatible observables which are described by three operators which commute with each other.

For example, we could specify the x and y coordinates and the z component of the momentum of a particle to define a quantum state. These observables are described by the operators

$$\hat{x} = x, \quad \hat{y} = y \quad \text{and} \quad \hat{p}_z = -i\hbar\frac{\partial}{\partial z}.$$

It is easy to show that any two of these operators commute. For example

$$\hat{x}\hat{p}_z \, \Psi = x\left(-i\hbar\frac{\partial}{\partial z}\right)\Psi = -i\hbar x\frac{\partial}{\partial z}\Psi$$

and

$$\hat{p}_z\hat{x} \, \Psi = \left(-i\hbar\frac{\partial}{\partial z}\right)x\Psi = -i\hbar x\frac{\partial}{\partial z}\Psi.$$

In fact, we have three commuting operators,

$$[\hat{x}, \hat{y}] = [\hat{x}, \hat{p}_z] = [\hat{y}, \hat{p}_z] = 0,$$

and simultaneous eigenfunctions of the form

$$\psi_{x'y'p'_z}(x, y, z) = \delta(x - x')\delta(y - y')\frac{1}{\sqrt{2\pi\hbar}}\,e^{-ip'_z z}.$$

Moreover, any wave function $\Psi(x, y, z, t)$ can be expressed as linear superposition of these eigenfunctions as follows:

$$\Psi(x, y, z, t) = \int_{-\infty}^{+\infty} dx' \int_{-\infty}^{+\infty} dy' \int_{-\infty}^{+\infty} dp'_z\ c(x', y', p'_z, t)\psi_{x'y'p'_z}(x, y, z).$$

In this expression $c(x', y', p'_z, t)$ is a probability amplitude for three compatible observables. In fact, the probability of finding the particle at time t localized between x' and $x' + dx'$ and between y' and $y' + dy'$, and with momentum in the z direction between p'_z and $p'_z + dp'_z$, is $|c(x', y', p'_z, t)|^2\,dx'\,dy'\,dp'_z$.

This example has illustrated the general procedure of defining a quantum state of a particle moving in three dimensions by specifying a set of three compatible observables. This procedure will be used in Chapter 9 when we construct stationary states of the hydrogen atom by specifying the energy, the magnitude of the orbital angular momentum and the z component of the orbital angular momentum.

7.5   CONSTANTS OF MOTION

Observables that are compatible with the energy observable have a particular physical significance. They are *constants of the motion*. To explain the significance of this statement we consider the expectation value for an observable A for a particle with wave function Ψ,

$$\langle A(t) \rangle = \int \Psi^* \hat{A}\ \Psi\ d^3\mathbf{r}. \tag{7.20}$$

In general, the expectation value $\langle A(t) \rangle$ will vary with time as the wave function $\Psi(\mathbf{r}, t)$ ebbs and flows in accord with the Schrödinger equation

$$i\hbar \frac{\partial \Psi}{\partial t} = \hat{H}\Psi. \tag{7.21}$$

We can find the rate of change of $\langle A(t) \rangle$ by differentiating both sides of Eq. (7.20). Using the rules for differentiating a product of functions, we obtain[2]

$$\frac{\mathrm{d}\langle A \rangle}{\mathrm{d}t} = \int \frac{\partial \Psi^*}{\partial t} \hat{A} \, \Psi \, \mathrm{d}^3\mathbf{r} + \int \Psi^* \, \hat{A} \, \frac{\partial \Psi}{\partial t} \, \mathrm{d}^3\mathbf{r}.$$

If we use the Schrödinger equation (7.21) and the complex conjugate of this equation, we find

$$\frac{\mathrm{d}\langle A \rangle}{\mathrm{d}t} = -\frac{1}{i\hbar} \int (\hat{H}\Psi^*) \, \hat{A} \, \Psi \, \mathrm{d}^3\mathbf{r} + \frac{1}{i\hbar} \int \Psi^* \, \hat{A} \, (\hat{H}\Psi) \, \mathrm{d}^3\mathbf{r}.$$

Because the Hamiltonian $\hat{H}$, like any other operator for an observable in quantum mechanics, is a Hermitian operator, we can use Eq. (7.2) to show that

$$\int (\hat{H}\Psi^*) \, \hat{A} \, \Psi \, \mathrm{d}^3\mathbf{r} = \int \Psi^* \, \hat{H}\hat{A} \, \Psi \, \mathrm{d}^3\mathbf{r}$$

and rewrite the expression for the rate of change of $\langle A(t) \rangle$ as

$$\frac{\mathrm{d}\langle A \rangle}{\mathrm{d}t} = \frac{1}{i\hbar} \int \Psi^* \, [\hat{A}, \hat{H}] \, \Psi \, \mathrm{d}^3\mathbf{r}, \tag{7.22}$$

where $[\hat{A}, \hat{H}]$ is a commutator. This equation can be used to determine the time-dependence of the expectation value of any observable.

For an observable A which is compatible with the energy, the commutator $[\hat{A}, \hat{H}]$ is zero and Eq. (7.22) gives

$$\frac{\mathrm{d}\langle A \rangle}{\mathrm{d}t} = 0.$$

Such an observable is called a constant of motion because its expectation value does not change as the wave function evolves with time.

These ideas can be illustrated by considering a particle with the Hamiltonian

$$\hat{H} = -\frac{\hbar^2}{2m}\nabla^2 + V(r),$$

where $V(r)$ is a potential energy which only depends on the distance r of the particle from a fixed origin. For this Hamiltonian, it is easy to show that

---

[2] We are assuming that the operator $\hat{A}$ does not depend on time. This assumption is possible for any isolated system.

$$[\hat{\mathbf{r}}, \hat{H}] \neq 0, \quad [\hat{\mathbf{p}}, \hat{H}] \neq 0, \quad \text{and} \quad [\hat{\mathbf{L}}, \hat{H}] = 0.$$

These equations imply that the position and the momentum are not constants of motion but that the orbital angular momentum is a constant of motion. In fact, the constants of motion of a system are determined by the symmetry properties of its Hamiltonian. In this example, the Hamiltonian has rotational symmetry and this symmetry implies that the orbital angular momentum is a constant of motion.

PROBLEMS 7

1. Consider a particle moving in a one-dimensional potential energy field $V(x)$. Show that the operators describing the position, momentum and energy of the particle satisfy the following mathematical relations:

$$[\hat{x}, \hat{p}] \neq 0, \quad [\hat{x}, \hat{H}] \neq 0, \quad \text{and} \quad [\hat{p}, \hat{H}] \neq 0.$$

What is the physical significance of these mathematical relations?

2. (a) Write down the kinetic energy operator $\hat{T}$ and momentum operator $\hat{p}$ for a particle of mass m moving along the x axis.

   (b) Show that $\hat{T}$ and $\hat{p}$ satisfy the commutation relation

   $$[\hat{p}, \hat{T}] = 0.$$

   Explain the physical significance of this result.

   (c) Show that

   $$\psi(x) = A \cos kx$$

   is an eigenfunction of the kinetic operator $\hat{T}$ but not a eigenfunction of $\hat{p}$.

   (d) Are there wave functions which are simultaneously eigenfunctions of $\hat{T}$ and $\hat{p}$? If so, write one down.

3. The canonical commutation relations for a particle moving in three dimensions are

$$[\hat{x}, \hat{p}_x] = i\hbar, \quad [\hat{y}, \hat{p}_y] = i\hbar, \quad \text{and} \quad [\hat{z}, \hat{p}_z] = i\hbar,$$

and all other commutators involving $\hat{x}$, $\hat{p}_x$, $\hat{y}$, $\hat{p}_y$, $\hat{z}$, $\hat{p}_z$ are zero. These relations can be used to show that the operators for the orbital angular momentum obey the following commutation relations:

$$[\hat{L}_x, \hat{L}_y] = i\hbar\hat{L}_z, \quad [\hat{L}_y, \hat{L}_z] = i\hbar\hat{L}_x, \quad \text{and} \quad [\hat{L}_z, \hat{L}_x] = i\hbar\hat{L}_y.$$

(a) Using

$$\hat{L}_x = \hat{y}\hat{p}_z - \hat{z}\hat{p}_y \quad \text{and} \quad \hat{L}_y = \hat{z}\hat{p}_x - \hat{x}\hat{p}_z,$$

verify that

$$[\hat{L}_x, \hat{L}_y] = [\hat{y}\hat{p}_z, \hat{z}\hat{p}_x] + [\hat{z}\hat{p}_y, \hat{x}\hat{p}_z].$$

(b) Using the commutation relations

$$[\hat{z}, \hat{p}_z] = i\hbar \quad \text{and} \quad [\hat{y}, \hat{p}_y] = i\hbar,$$

verify that

$$[\hat{L}_x, \hat{L}_y] = i\hbar\hat{L}_z.$$

What is the physical significance of this result?

(c) Using

$$[\hat{L}_x, \hat{L}_y] = i\hbar\hat{L}_z, \quad [\hat{L}_y, \hat{L}_z] = i\hbar\hat{L}_x, \quad \text{and} \quad [\hat{L}_z, \hat{L}_x] = i\hbar\hat{L}_y,$$

verify that

$$[\hat{L}^2, \hat{L}_z] = 0$$

where

$$\hat{L}^2 = \hat{L}_x^2 + \hat{L}_y^2 + \hat{L}_z^2.$$

What is the physical significance of this result?

(Hint: For any two operators $\hat{A}$ and $\hat{B}$,

$$[\hat{A}^2, \hat{B}] = \hat{A}\hat{A}\hat{B} - \hat{B}\hat{A}\hat{A} = \hat{A}\hat{A}\hat{B} - \hat{A}\hat{B}\hat{A} - \hat{B}\hat{A}\hat{A} + \hat{A}\hat{B}\hat{A}$$

implies that

$$[\hat{A}^2, \hat{B}] = \hat{A}[\hat{A}, \hat{B}] + [\hat{A}, \hat{B}]\hat{A}. \quad)$$

4. In this problem and in problems 5 and 6 we shall consider some general mathematical properties of the operators which describe observables in quantum mechanics. To keep the mathematics as simple as possible we shall only consider a particle moving along the x axis.

   In general, an operator $\hat{A}$ describing an observable A is a *Hermitian operator* which means that it obeys the condition

$$\int_{-\infty}^{+\infty} \Psi_1^* \, \hat{A} \, \Psi_2 \, dx = \int_{-\infty}^{+\infty} (\hat{A}\Psi_1)^* \, \Psi_2 \, dx,$$

where Ψ_1 and Ψ_2 are any two wave functions; the brackets in the term $(\hat{A}\Psi_1)^*$ mean that the operator $\hat{A}$ only acts on the wave function Ψ_1 and the complex conjugate of the result is taken.

   By integrating by parts and by assuming that the wave functions go to zero at infinity, verify that the momentum operator $\hat{p} = -i\hbar\partial/\partial x$ is a Hermitian operator.

5. Consider a particle with wave function Ψ.

   (a) Bearing in mind that an observable A is described by a Hermitian operator, show that its expectation value

$$\langle A \rangle = \int_{-\infty}^{+\infty} \Psi^* \, \hat{A} \, \Psi \, dx$$

   is real.

   (b) Show that the expectation value of A^2 is given by

$$\langle A^2 \rangle = \int_{-\infty}^{+\infty} (\hat{A}\Psi)^* \, (\hat{A}\Psi) \, dx.$$

   (c) Show for two Hermitian operators $\hat{A}$ and $\hat{B}$ that

$$\int_{-\infty}^{+\infty} \Psi^* \, \hat{A}\hat{B} \, \Psi \, dx = \left(\int_{-\infty}^{+\infty} \Psi^* \, \hat{B}\hat{A} \, \Psi \, dx \right)^*.$$

   Hence, show that

$$\int_{-\infty}^{+\infty} \Psi^* \, (\hat{A}\hat{B} + \hat{B}\hat{A}) \, \Psi \, dx$$

is real and that

$$\int_{-\infty}^{+\infty} \Psi^* \, (\hat{A}\hat{B} - \hat{B}\hat{A}) \, \Psi \, dx$$

is imaginary.

6. In this problem we shall derive an inequality called the *Schwarz inequality* and a related inequality which is used to derive the Heisenberg uncertainty relation.

Let $\alpha(x)$ and $\beta(x)$ be complex functions of x which give finite values for the integrals

$$\int_{-\infty}^{+\infty} |\alpha|^2 \, dx, \quad \int_{-\infty}^{+\infty} |\beta|^2 \, dx, \quad \text{and} \quad \int_{-\infty}^{+\infty} \alpha^*\beta \, dx,$$

and let $\phi(x)$ be a complex function given by

$$\phi(x) = \alpha(x) + \lambda\beta(x)$$

where λ is a complex number. Because

$$\int_{-\infty}^{+\infty} |\phi|^2 \, dx \geq 0,$$

we have

$$\int_{-\infty}^{+\infty} |\alpha|^2 \, dx + \lambda \int_{-\infty}^{+\infty} \alpha^*\beta \, dx + \lambda^* \int_{-\infty}^{+\infty} \beta^*\alpha \, dx + \lambda^*\lambda \int_{-\infty}^{+\infty} |\beta|^2 \, dx \geq 0.$$

Because this inequality is valid for any value of λ, it is valid when λ is given by

$$\lambda \int_{-\infty}^{+\infty} |\beta|^2 \, dx = -\int_{-\infty}^{+\infty} \beta^*\alpha \, dx.$$

(a) Verify the Schwarz inequality

$$\int_{-\infty}^{+\infty} |\alpha|^2 \, dx \int_{-\infty}^{+\infty} |\beta|^2 \, dx \geq \left|\int_{-\infty}^{+\infty} \alpha^*\beta \, dx\right|^2.$$

(b) Identify $\hat{A}\Psi$ with the function $\alpha(x)$ and $\hat{B}\Psi$ with the function $\beta(x)$ and use problem 5(b) to show that

$$\langle A^2 \rangle \langle B^2 \rangle \geq \left| \int_{-\infty}^{+\infty} \Psi^* \, \hat{A} \, \hat{B} \, \Psi \, dx \right|^2 .$$

(c) Now use problem 5(c) to show that $\langle A^2 \rangle \langle B^2 \rangle$ is greater than or equal to

$$\left| \int_{-\infty}^{+\infty} \Psi^* \left(\frac{\hat{A} \hat{B} + \hat{B} \hat{A}}{2} \right) \Psi \, dx \right|^2 \quad + \quad \left| \int_{-\infty}^{+\infty} \Psi^* \left(\frac{\hat{A} \hat{B} - \hat{B} \hat{A}}{2} \right) \Psi \, dx \right|^2 .$$

7. (a) A particle has a Hamiltonian of the form

$$\hat{H} = -\frac{\hbar^2}{2m} \left(\frac{\partial^2}{\partial x^2} + \frac{\partial^2}{\partial y^2} + \frac{\partial^2}{\partial z^2} \right) + V(x, y, z).$$

What symmetry property must be satisfied by the potential energy field $V(x, y, z)$ in order that the x component of the momentum of the particle is a constant of motion?

(b) A particle has a Hamiltonian of the form

$$\hat{H} = -\frac{\hbar^2}{2m} \left(\frac{\partial^2}{\partial r^2} + \frac{1}{r} \frac{\partial}{\partial r} + \frac{1}{r^2} \frac{\partial^2}{\partial \phi^2} + \frac{\partial^2}{\partial z^2} \right) + V(r, \phi, z),$$

where (r, ϕ, z) are cylindrical coordinates.

What symmetry properties must be satisfied by the potential energy field $V(r, \phi, z)$ in order that the z component of the momentum and the z component of the orbital angular momentum, two observables described by the operators

$$\hat{p}_z = -i\hbar \frac{\partial}{\partial z} \quad \text{and} \quad \hat{L}_z = -i\hbar \frac{\partial}{\partial \phi},$$

are constants of the motion?

8. In this question you are asked to derive the *virial theorem* for a particle with Hamiltonian $\hat{H} = \hat{T} + \hat{V}$, with a kinetic energy operator given by

$$\hat{T} = -\frac{\hbar^2 \nabla^2}{2m}$$

and a potential energy operator given by

$$\hat{V} = V(r).$$

(a) Show that

$$[\hat{\mathbf{r}} \cdot \hat{\mathbf{p}}, \hat{T}] = \frac{i\hbar}{m} \hat{\mathbf{p}}^2 \quad \text{and that} \quad [\hat{\mathbf{r}} \cdot \hat{\mathbf{p}}, \hat{V}] = -i\hbar r \frac{\mathrm{d}V}{\mathrm{d}r}.$$

(b) Consider $\psi_E(\mathbf{r})$, an eigenfunction of $\hat{H}$ with eigenvalue E. Bearing in mind that $\hat{H}$ is a Hermitian operator, show that

$$\int \psi_E^* \, [\hat{\mathbf{r}} \cdot \hat{\mathbf{p}}, \hat{H}] \, \psi_E \, \mathrm{d}^3\mathbf{r} = 0.$$

Hence show that

$$2 \int \psi_E^* \, \hat{T} \, \psi_E \, \mathrm{d}^3\mathbf{r} = \int \psi_E^* \, r \frac{\mathrm{d}V}{\mathrm{d}r} \, \psi_E \, \mathrm{d}^3\mathbf{r}$$

which is a statement of the virial theorem.

(c) Show that the expectation values for the kinetic and the potential energies of a particle in a state with definite energy are related by

$$\langle T \rangle = \langle V \rangle \quad \text{if the potential is} \quad V(r) = \tfrac{1}{2} m \omega^2 r^2,$$

and by

$$2\langle T \rangle = \langle V \rangle \quad \text{if the potential is} \quad V(r) = -\frac{e^2}{4\pi\epsilon_0 r}.$$

8

Angular momentum

Planck's constant has the units of angular momentum. This suggests that h, or $\hbar = h/2\pi$, may be the fundamental unit for angular momentum. It also suggests that angular momentum may be a fundamental observable in quantum physics. Indeed, there are point-like quantum particles which have an intrinsic angular momentum called *spin*. The spin of a point particle cannot be related to orbital motion of constituent parts; it is a fundamental property which has no analogue in classical physics.

The nitty-gritty of spin and orbital angular momentum forms a major part of advanced books on quantum mechanics. In this chapter, we shall only set out and illustrate the most important aspects of this demanding topic. We shall begin by considering the basic properties of angular momentum and then describe how these properties may be revealed by the interaction of magnetic moments with magnetic fields. Finally, we shall describe how the angular shape of a wave function is related to the orbital angular momentum of the particle described by this wave function.

8.1 ANGULAR MOMENTUM BASICS

Quantum particles may possess an orbital angular momentum and an intrinsic angular momentum, called *spin*. In appropriate circumstances, the orbital angular momentum resembles the orbital angular momentum of a classical particle; it is a vector, with direction and magnitude describing the inertia of angular motion. In contrast, spin angular momentum does not have a classical manifestation; it is a fundamental quantum property which bears little resemblance to a rotating classical object.

The most important property of angular momentum in quantum mechanics is that the outcome of measurement is at best a fuzzy vector, a vector with two defining properties: a definite magnitude and a definite value for just *one* of its three Cartesian components. Accordingly, an angular momentum in quantum physics may be specified using two quantum numbers. Normally, l and m_l are used to describe orbital angular momentum, s and m_s are used for spin angular momentum, and the quantum numbers j and m_j are used when the angular momentum arises from a combination of spin and orbital angular momentum, and when a general angular momentum is being described.[1]

The only possible precise values for the magnitude of orbital angular momentum are given by

$$L = \sqrt{l(l+1)}\hbar, \quad \text{where} \quad l = 0, 1, 2, 3, \ldots. \tag{8.1}$$

When the magnitude is fixed by the quantum number l, the orbital angular momentum in any given direction may have $2l + 1$ possible values between $-l\hbar$ and $+l\hbar$. For example, if we choose to measure the orbital angular momentum in the z direction, there are $2l + 1$ possible outcomes given by

$$L_z = m_l \hbar, \quad \text{where} \quad m_l = \begin{cases} +l \\ +(l-1) \\ +(l-2) \\ \vdots \\ -(l-2) \\ -(l-1) \\ -l. \end{cases} \tag{8.2}$$

But when this is done, the orbital angular momentum in the x and y directions are uncertain, in the sense that, if we choose to measure the x or y component, the possible outcomes will have quantized values somewhere in the range $-l\hbar$ to $+l\hbar$. We shall check the validity of all these general statements about orbital angular momentum in Section 8.3. Readers should note for future reference that the classification of atomic spectra has led to the spectroscopic notation in which the letters s, p, d, f, and g are used to label quantum states with $l = 0, l = 1, l = 2, l = 3$ and $l = 4$. Accordingly, states with $l = 0$ are called s-states, states with $l = 1$ are called p-states, and so on.

The quantum numbers s and m_s are usually used when the angular momentum is solely due to spin. A particle is said to have spin s if the magnitude of the spin angular momentum is $S = \sqrt{s(s+1)}\hbar$ and if the z components are given by

$$S_z = m_s \hbar, \quad \text{where} \quad m_s = \begin{cases} +s \\ +(s-1) \\ \vdots \\ -(s-1) \\ -s. \end{cases} \tag{8.3}$$

---

[1] In Chapter 11 we shall use capital letters, L, S and J, for the angular momentum quantum numbers of two or more electrons, but in this chapter these letters will denote the magnitude of an angular momentum.

For example, the W boson is a spin-one particle with $s = 1$ and $m_s = +1, 0, -1$ and the electron is a spin-half particle with $s = \frac{1}{2}$ and $m_s = \pm\frac{1}{2}$. Thus, spin angular momentum can be integer, like orbital angular momentum, but it can also be half-integer.

Orbital and spin angular momenta may be combined to give a total angular momentum with magnitude and z component given by

$$J = \sqrt{j(j+1)}\hbar \quad \text{and} \quad J_z = m_j\hbar, \tag{8.4}$$

where, in general, the quantum numbers j and m_j may take on integer and half-integer values given by

$$j = 0, \frac{1}{2}, 1, \frac{3}{2}, 2, \dots\dots \quad \text{and} \quad m_j = \begin{cases} +j \\ +(j-1) \\ \vdots \\ -(j-1) \\ -j. \end{cases} \tag{8.5}$$

The actual values of the quantum number j depend on the orbital and spin angular momenta being combined. It can be shown that, when an orbital angular momentum with quantum number l is combined with a spin with quantum number s, several total angular momenta may arise with quantum numbers

$$j = l+s, l+s-1, \dots |l-s|. \tag{8.6}$$

For example, we can have $j = \frac{3}{2}$ and $\frac{1}{2}$ when $l = 1$ and $s = \frac{1}{2}$, and we can have $j = 2, 1$ and 0 when $l = 1$ and $s = 1$. We note that, in general, two angular momenta with quantum numbers j_1 and j_2 may be combined to give an angular momentum with quantum number j which can take on the values

$$j = j_1 + j_2, j_1 + j_2 - 1, \dots , |j_1 - j_2|.$$

Earlier we referred to an angular momentum defined by two quantum numbers as a fuzzy vector. The fuzziness arises because, when one of its Cartesian components is sharply defined, the other two components are uncertain but quantized when measured. In view of the uncertainties we have already encountered in position, momentum and energy, uncertain angular momentum should not be a surprise. Indeed, the uncertainty in orbital angular momentum can be directly traced to the uncertainties in the position and momentum of a particle, as indicated in problem 3 at the end of Chapter 7. But it is surprising that angular momentum in any given direction can only equal an integer or half-integer multiple of $\hbar$. We shall see how this surprising property may be confirmed experimentally in the next section.[2]

8.2 MAGNETIC MOMENTS

In this section we shall consider magnetic moments and then describe how the interaction of a magnetic moment with a magnetic field can reveal the properties of an angular momentum.

Classical magnets

The simplest magnetic moment in classical physics consists of an orbiting charged particle. This magnetic moment is directly proportional to the orbital angular momentum and it is given by

$$\boldsymbol{\mu} = \frac{q}{2m}\mathbf{L}, \tag{8.7}$$

where q is the charge and m is the mass of the orbiting particle.

We can check the validity of this relation by considering a particle moving in a circular orbit of radius r with speed v, as shown in Fig. 8.1. Such a particle gives rise to a circulating electrical current, I, and hence to a magnetic moment with magnitude IA, where A is the area of the orbit. Because the current is equal to the charge q divided by the period of the orbit $2\pi r/v$, we have

$$\mu = \frac{qv}{2\pi r}\,\pi r^2 = \frac{qrv}{2},$$

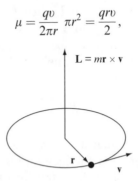

Fig. 8.1 A particle with charge q and mass m moving in a circular orbit with orbital angular momentum $\mathbf{L} = m\mathbf{r} \times \mathbf{v}$ gives rise to a magnetic moment $\boldsymbol{\mu} = q\mathbf{L}/2m$.

---

[2] More advanced texts show that the operators for angular momentum also generate rotations. The geometrical properties of rotations in three dimensions imply that the angular momentum operators obey commutation relations which require angular momentum to be quantized in integer and half-integer multiples of $\hbar$.

which may be rewritten using the angular momentum $L = mrv$ as

$$\mu = \frac{q}{2m}L.$$

The vector version of this equation, Eq. (8.7), follows because the directions of the magnetic moment and of the orbital angular momentum are both perpendicular to the plane of the orbit.

Quantum magnets

In quantum physics, magnetic moments are also proportional to angular momenta, but they are now at best fuzzy vectors, with precise values for their magnitude and for one Cartesian component. We shall illustrate the relation between quantum magnets and angular momenta by considering the magnetic properties of electrons, atoms, protons and neutrons.

The magnetic properties of an electron arise from a spin with magnitude and z component given by

$$S = \sqrt{\tfrac{1}{2}\left(\tfrac{1}{2}+1\right)}\,\hbar \quad \text{and} \quad S_z = m_s\hbar,$$

and from an orbital angular momentum with magnitude and z component given by

$$L = \sqrt{l(l+1)}\,\hbar \quad \text{and} \quad L_z = m_l\hbar.$$

The associated magnetic moments are given by formulae similar to Eq. (8.7), but with minor modifications. In particular, the z components of the magnetic moments due to electron spin are

$$\mu_z^{(\text{Spin})} = -2\frac{e}{2m_e}S_z = -2\frac{e\hbar}{2m_e}m_s \qquad (8.8)$$

and those due to orbital angular momentum are

$$\mu_z^{(\text{Orbital})} = -\frac{e}{2m_e}L_z = -\frac{e\hbar}{2m_e}m_l, \qquad (8.9)$$

where m_e is the mass and $-e$ is the charge of an electron. We note that electron spin angular momentum is twice as magnetic as orbital angular momentum; the additional factor of 2 in Eq. (8.8) is explained by the Dirac equation, a wave equation for relativistic, point-like quantum particles with spin half.

The magnetic moment of an atom arises from the combined spin and orbital angular momentum of the constituent electrons which can be described by quantum numbers j and m_j. In a weak magnetic field, the z component of the magnetic moment of an atom is

$$\mu_z^{(\text{Atom})} = -g\frac{e}{2m_e}J_z = -g\frac{e\hbar}{2m_e}m_j, \tag{8.10}$$

where $J_z = m_j\hbar$ is the z component of the angular momentum and g is a numerical factor called the Landé g-factor. In fact, the Landé g-factor for an atomic state with quantum numbers j, l and s is

$$g = 1 + \frac{j(j+1) - l(l+1) + s(s+1)}{2j(j+1)}.$$

The Landé g-factor has the value $g = 2$ if the sole source of magnetism in the atom is due to electron spin, and the value $g = 1$ if due to orbital angular momentum. Because m_j can take on $2j + 1$ values between $-j$ and $+j$, the magnetic moment of the atom has $2j + 1$ components.

Equations (8.8), (8.9), and (8.10) indicate that the natural unit for magnetic moments associated with electrons is

$$\mu_B = \frac{e\hbar}{2m_e} = 9.274 \times 10^{-24} \text{ J T}^{-1}. \tag{8.11}$$

This fundamental constant is called the *Bohr magneton*.

Protons and neutrons, unlike electrons, are composite objects containing quarks and gluons. These constituents give rise to angular momenta with quantum numbers $j = \frac{1}{2}$ and $m_j = \pm\frac{1}{2}$. The z components of the associated magnetic moments are

$$\mu_z^{(\text{Proton})} = 2.79\frac{e\hbar}{2m_p}m_j \quad \text{and} \quad \mu_z^{(\text{Neutron})} = -1.95\frac{e\hbar}{2m_p}m_j, \tag{8.12}$$

where m_p is the mass of the proton. We note that the natural unit for these magnetic moments, and also for the magnetic moments of nuclei containing protons and neutrons, is

$$\mu_N = \frac{e\hbar}{2m_p} = 5.05 \times 10^{-27} \text{ J T}^{-1}. \tag{8.13}$$

This unit is called the *nuclear magneton*.

Magnetic energies and the Stern–Gerlach experiment

When a classical magnetic moment $\boldsymbol{\mu}$ is placed in a magnetic field **B**, it has an energy of orientation given by

$$E_{mag} = -\boldsymbol{\mu} \cdot \mathbf{B}. \qquad (8.14)$$

If we choose the direction of the magnetic field to be the z direction, we have

$$E_{mag} = -\mu_z B, \qquad (8.15)$$

where μ_z is the z component of the magnetic moment, which can take on any value between $+\mu$ and $-\mu$. Hence, for a classical magnet, there is a continuum of energies of orientation between $-\mu B$ and $+\mu B$.

In marked contrast, the energy of orientation of a quantum magnet in a magnetic field is quantized. For a field B in the z direction, this energy is given by $-\mu_z B$, where μ_z is now the quantized z component of the magnetic moment. For example, we can use Eq. (8.10) to show that the magnetic energy of an atom in an atomic state with angular quantum numbers j and m_j is

$$E_{mag} = m_j \, g\mu_B B,$$

where μ_B is the Bohr magneton and g the Landé g-factor. Thus, for a given value of j, there are $2j + 1$ magnetic energy levels given by

$$E_{mag} = \begin{cases} +j \, g\mu_B B \\ +(j-1) \, g\mu_B B \\ \vdots \\ -(j-1) \, g\mu_B B \\ -j \, g\mu_B B. \end{cases} \qquad (8.16)$$

When $j = \frac{1}{2}$ there are two energy levels, when $j = 1$ there are three energy levels, when $j = \frac{3}{2}$ there are four energy levels, and so on, as shown in Fig. 8.2.

Indirect evidence for atomic magnetic energy levels is provided by observing the effect of a magnetic field on spectral lines. The magnetic field splits atomic energy levels with a given j into $2j + 1$ magnetic energy levels with different values for m_j, and radiative transitions between states with different values of j now give rise to several closely spaced spectral lines instead of one. This effect is called the *Zeeman effect*.

However, direct evidence for the quantization of magnetic energies is provided by a Stern–Gerlach experiment. In this experiment individual atoms pass through a non-uniform magnetic field which separates out the atoms according the value of their magnetic moment in a given direction.

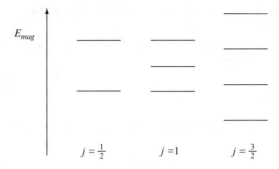

Fig. 8.2 The energy levels in a magnetic field of an atom in states with angular momentum quantum numbers $j = \frac{1}{2}$, 1 and $\frac{3}{2}$. The spacing between levels is given by $g\mu_B B$ where B is the strength of the magnetic field, μ_B is the Bohr magneton and g is a Landé g-factor, a constant which depends on the spin and orbital angular momentum quantum numbers of the atomic state.

The main features of a Stern–Gerlach experiment are illustrated in Fig. 8.3. A beam of atoms is passed through a magnetic field produced by specially shaped poles of an electromagnet. The direction of the magnetic field is largely in one direction, the z direction say, but its strength, $B(x, y, z)$, increases markedly as z increases. In this field, each atom acquires an energy

$$E_{mag}(x, y, z) = -\mu_z B(x, y, z)$$

which depends upon the z component of its magnetic moment μ_z and on the location in the field. Because this magnetic energy varies strongly with z, the atom is deflected by a force in the z direction which is given by

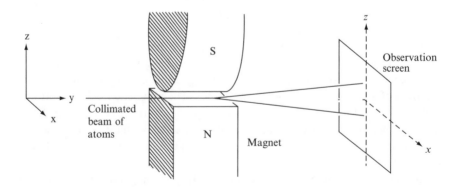

Fig. 8.3 The Stern–Gerlach experiment in which atoms pass through a non-uniform magnetic field which separates out atoms according to the value of the magnetic moment in the direction of maximum non-uniformity of the field.

$$F = -\frac{\partial E_{mag}}{\partial z} = \mu_z \frac{\partial B}{\partial z}.$$

If the z component of the magnetic moment could take on any value between $+\mu$ and $-\mu$, the atomic beam would be smeared out as atoms are dragged up and down by varying amounts. But for real atoms in states with quantum number j, the z component of the magnetic moment can only take on $2j + 1$ discrete values and a beam of such atoms is split into $2j + 1$ separate beams.

In their original experiment, Stern and Gerlach discovered that a beam of silver atoms, in their ground state, is split into two separate beams. This implies that a measured Cartesian component of the magnetic moment of a silver atom in its ground state can only take on two possible values and that the angular momentum quantum numbers for the atom are $j = \frac{1}{2}$ and $m_j = \pm\frac{1}{2}$. They also showed, by measuring the separation between the two beams of atoms emerging from the electromagnet, that the magnitude of the magnetic moment of a silver atom is of the order of a Bohr magneton.

8.3   ORBITAL ANGULAR MOMENTUM

In this section we shall remind the reader of the definition of orbital angular momentum in classical physics, introduce the operators which describe orbital angular momentum in quantum physics, and then consider how the angular shapes of wave functions are related to orbital angular momentum properties. In so doing, we shall confirm some of the general statements made about orbital angular momentum in Section 8.1.

Classical orbital angular momentum

Consider a particle at time t with vector position and momentum

$$\mathbf{r} = (x, y, z) \quad \text{and} \quad \mathbf{p} = (p_x, p_y, p_z).$$

The orbital angular momentum about the origin of coordinates is given by the vector product

$$\mathbf{L} = \mathbf{r} \times \mathbf{p},$$

which is a vector with three Cartesian components,

$$L_x = yp_z - zp_y, \quad L_y = zp_x - xp_z, \quad L_z = xp_y - yp_x,$$

and a magnitude given by

$$|\mathbf{L}| = \sqrt{L_x^2 + L_y^2 + L_z^2}.$$

Quantum orbital angular momentum

Orbital angular momentum in quantum physics is described by the operator

$$\hat{\mathbf{L}} = \hat{\mathbf{r}} \times \hat{\mathbf{p}} = -i\hbar \mathbf{r} \times \nabla. \qquad (8.17)$$

This a vector operator with three Cartesian components,

$$\hat{L}_x = -i\hbar\left(y\frac{\partial}{\partial z} - z\frac{\partial}{\partial y}\right), \quad \hat{L}_y = -i\hbar\left(z\frac{\partial}{\partial x} - x\frac{\partial}{\partial z}\right),$$

$$\hat{L}_z = -i\hbar\left(x\frac{\partial}{\partial y} - y\frac{\partial}{\partial x}\right),$$

that act on wave functions representing possible quantum states of a particle.[3]

When the wave function $\Psi(\mathbf{r}, t)$ is known, expectation values for orbital angular momentum may be calculated. For example, the integrals

$$\langle L_x \rangle = \int \Psi^* \, \hat{L}_x \, \Psi \, \mathrm{d}^3\mathbf{r} \quad \text{and} \quad \langle L_x^2 \rangle = \int \Psi^* \, \hat{L}_x^2 \, \Psi \, \mathrm{d}^3\mathbf{r}$$

give the expectation values for the x component and the square of the x component of the orbital angular momentum. And when the wave function is an eigenfunction of $\hat{L}_x$ with eigenvalue L_x, i.e. when

$$\hat{L}_x\Psi(\mathbf{r}, t) = L_x\Psi(\mathbf{r}, t),$$

we can follow the procedure outlined in Section 4.3 and show that $\langle L_x \rangle = L_x$ and that $\langle L_x^2 \rangle = L_x^2$. This implies that the uncertainty $\Delta L_x = \sqrt{\langle L_x^2 \rangle - \langle L_x \rangle^2}$ is zero and that the eigenfunction represents a quantum state with a precise value for the x component of the orbital angular momentum given by the eigenvalue L_x.

Angular shape of wave functions

The wave function of a particle can have an infinite variety of angular shapes. But any wave function can be expressed in terms of basis wave functions with

---

[3] Spin angular momentum is usually described by an operator $\hat{\mathbf{S}} = (\hat{S}_x, \hat{S}_y, \hat{S}_z)$ which acts on a quantum state which includes a description of the spin properties of the particle. Spin operators and spin quantum states are usually represented by matrices.

simpler angular shapes. These basis wave functions are usually taken to be wave functions with specific orbital angular momentum properties. Accordingly, we shall consider some wave functions with simple angular dependence and deduce the orbital angular momentum properties of the particle they describe. The properties of the following wave functions will be explored: The spherically symmetric wave function given by

$$\psi_{(0,0)} = R(r), \tag{8.18}$$

where $R(r)$ is any well-behaved function of $r = \sqrt{x^2 + y^2 + z^2}$, and the wave functions

$$\psi_{(1,0)} = R(r)\frac{z}{r}, \quad \psi_{(1,+1)} = R(r)\frac{(x+iy)}{r}, \quad \psi_{(1,-1)} = R(r)\frac{(x-iy)}{r}. \tag{8.19}$$

The rationale for the labels (0,0), (1,0) and (1, ± 1) will become clear after we have determined the angular momentum properties of the states described by these wave functions.

The position probability densities for these wave functions,

$$|\psi_{(0,0)}|^2 = |R(r)|^2, \quad |\psi_{(1,0)}|^2 = |R(r)|^2\frac{z^2}{r^2} \quad \text{and} \quad |\psi_{(1,\pm1)}|^2 = |R(r)|^2\frac{(x^2+y^2)}{r^2},$$

are illustrated in Fig. 8.4. We note that a particle described by the wave function $\psi_{(0,0)}$ is equally likely to be found at any point on the surface of a sphere of radius r, whereas particular regions of the surface are more likely locations for a particle described by the wave functions $\psi_{(1,0)}$ and $\psi_{(1,\pm1)}$. For the wave function $\psi_{(1,0)}$ the North and South poles are more probable

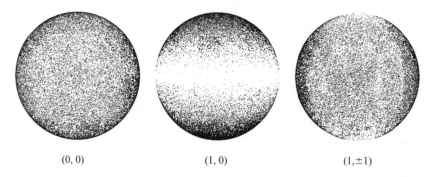

(0, 0)   (1, 0)   (1,±1)

Fig. 8.4 The position probability densities on the surface of a sphere for a particle with wave functions $\psi_{(0,0)}, \psi_{(1,0)}$ and $\psi_{(1,\pm1)}$ given by Eqs. (8.18) and (8.19). For future reference, these wave functions have orbital angular quantum numbers (l, m_l) equal to (0, 0), (1, 0) and (1, ± 1). (This figure was produced with the permission of Thomas D. York.)

locations, and for the wave functions $\psi_{(1,\pm1)}$ equatorial regions near $z = 0$ are more likely.

To find the orbital angular momentum properties of the particle described by wave functions Eqs. (8.18) and (8.19), we consider the action of the angular momentum operator given in Eq. (8.17) on these wave functions.

We first consider the action of the vector operator $\hat{\mathbf{L}}$ on the function $R(r)$. Using

$$\hat{\mathbf{L}} = -i\hbar \mathbf{r} \times \nabla \quad \text{and} \quad \nabla R(r) = \mathbf{e_r} \frac{dR}{dr},$$

where $\mathbf{e_r}$ is a unit vector in the direction of $\mathbf{r}$, we obtain

$$\hat{\mathbf{L}}\, R(r) = -i\hbar \mathbf{r} \times \nabla R(r) = -i\hbar \mathbf{r} \times \mathbf{e_r} \frac{dR}{dr}.$$

Because $\mathbf{r} \times \mathbf{e_r} = 0$, we deduce that $\hat{\mathbf{L}}R(r) = 0$. Hence, the spherically symmetric wave function $\psi_{(0,0)} = R(r)$ satisfies the three equations:

$$\hat{L}_x\psi_{(0,0)} = L_x\psi_{(0,0)}, \quad \text{with } L_x = 0,$$
$$\hat{L}_y\psi_{(0,0)} = L_y\psi_{(0,0)}, \quad \text{with } L_y = 0,$$
$$\hat{L}_z\psi_{(0,0)} = L_z\psi_{(0,0)}, \quad \text{with } L_z = 0.$$

It also satisfies the equation

$$\hat{L}^2\psi_{(0,0)} = L^2\psi_{(0,0)}, \quad \text{with} \quad L^2 = 0,$$

where

$$\hat{L}^2 = \hat{L}_x^2 + \hat{L}_y^2 + \hat{L}_z^2.$$

These equations show that any spherically symmetric wave function is a simultaneous eigenfunction of the operators which describe the magnitude and each of the three Cartesian components of the orbital angular momentum operator, and that in each case the eigenvalue is equal to zero. We conclude that all spherically symmetric wave functions describe a particle with zero orbital angular momentum.

We shall now consider the wave function

$$\psi_{(1,0)} = R(r)\frac{z}{r}$$

which describes a quantum particle that is more likely to be found near the North or South pole and not near the Equator, as shown in Fig. 8.4. To find the orbital angular properties of this particle, we evaluate the action of angular momentum operators on its wave function. Using the rules for the differentiation of a product, we obtain

$$\hat{\mathbf{L}}\left(\frac{R(r)}{r}z\right) = z\hat{\mathbf{L}}\left(\frac{R(r)}{r}\right) + \frac{R(r)}{r}\hat{\mathbf{L}}z = \frac{R(r)}{r}\hat{\mathbf{L}}z,$$

which implies that

$$\hat{L}_x\psi_{(1,0)} = \frac{R(r)}{r}\hat{L}_xz = -i\hbar\frac{R(r)}{r}\left(y\frac{\partial}{\partial z} - z\frac{\partial}{\partial y}\right)z = -i\hbar\frac{R(r)}{r}y,$$

$$\hat{L}_y\psi_{(1,0)} = \frac{R(r)}{r}\hat{L}_yz = -i\hbar\frac{R(r)}{r}\left(z\frac{\partial}{\partial x} - x\frac{\partial}{\partial z}\right)z = +i\hbar\frac{R(r)}{r}x,$$

$$\hat{L}_z\psi_{(1,0)} = \frac{R(r)}{r}\hat{L}_zz = -i\hbar\frac{R(r)}{r}\left(x\frac{\partial}{\partial y} - y\frac{\partial}{\partial x}\right)z = 0.$$

The first two equations show that the wave function $\psi_{(1,0)}$ is *not* an eigenfunction of $\hat{L}_x$ or of $\hat{L}_y$, but the third equation shows that $\psi_{(1,0)}$ is an eigenfunction of $\hat{L}_z$ with zero eigenvalue because

$$\hat{L}_z\psi_{(1,0)} = L_z\psi_{(1,0)}, \quad \text{with} \quad L_z = 0. \tag{8.20}$$

By evaluating terms like

$$\hat{L}_xz = -i\hbar y, \quad \hat{L}_xy = +i\hbar z \quad \text{and} \quad \hat{L}_x^2z = \hbar^2z,$$

it is also easy to show that

$$\hat{L}_x^2\psi_{(1,0)} = \hbar^2\psi_{(1,0)}.$$

Similarly we can easily show that

$$\hat{L}_y^2\psi_{(1,0)} = \hbar^2\psi_{(1,0)} \quad \text{and} \quad \hat{L}_z^2\psi_{(1,0)} = 0.$$

When we combine these results we find that

$$(\hat{L}_x^2 + \hat{L}_y^2 + \hat{L}_z^2)\psi_{(1,0)} = 2\hbar^2\psi_{(1,0)}. \tag{8.21}$$

Equations (8.20) and (8.21) show that the wave function $\psi_{(1,0)}$ is a simultaneous eigenfunction of the operators $\hat{L}^2 = \hat{L}_x^2 + \hat{L}_y^2 + \hat{L}_z^2$ and $\hat{L}_z$ with eigenvalues

$L^2 = 2\hbar^2$ and $L_z = 0$. Therefore, it describes a particle with a precise magnitude $L = \sqrt{2}\hbar$ and precise z component $L_z = 0$, but its orbital angular momentum in the x and y directions are uncertain.

Clearly, we can construct other wave functions with similar properties. For example, if we replace z in the expression for $\psi_{(1,0)}$ by x or by y, we obtain the wave functions

$$\psi'_{(1,0)} = R(r)\frac{x}{r} \quad \text{and} \quad \psi''_{(1,0)} = R(r)\frac{y}{r}. \tag{8.22}$$

Both these wave functions describe a particle with an orbital angular momentum of magnitude $L = \sqrt{2}\hbar$; but for $\psi'_{(1,0)}$ the x component is zero and the y and z components are uncertain, and for $\psi''_{(1,0)}$ the y component is zero and the z and x components are uncertain.

We shall finally consider the wave functions

$$\psi_{(1,+1)} = R(r)\frac{(x+iy)}{r} \quad \text{and} \quad \psi_{(1,-1)} = R(r)\frac{(x-iy)}{r},$$

both of which describe a quantum particle which is more likely to be found near the Equator and never at the North or South poles, as shown in Fig. 8.4. By evaluating the action of the angular momentum operators on the functions $x \pm iy$, it is easy to show that these wave functions are not eigenfunctions of $\hat{L}_x$ or of $\hat{L}_y$, but that they are both simultaneous eigenfunctions of $\hat{L}_z$ and $\hat{L}^2$. In fact,

$$\hat{L}_z\psi_{(1,+1)} = +\hbar\psi_{(1,+1)} \quad \text{and} \quad \hat{L}^2\psi_{(1,+1)} = 2\hbar^2\psi_{(1,+1)}$$

and

$$\hat{L}_z\psi_{(1,-1)} = -\hbar\psi_{(1,-1)} \quad \text{and} \quad \hat{L}^2\psi_{(1,-1)} = 2\hbar^2\psi_{(1,-1)}.$$

Thus, the wave function $\psi_{(1,+1)}$ describes a particle with $L_z = +\hbar$ and $L = \sqrt{2}\hbar$, and the wave function $\psi_{(1,-1)}$ describes a particle with $L_z = -\hbar$ and $L = \sqrt{2}\hbar$; in both cases, the x and y components of the orbital angular momentum are uncertain.

By exploring the properties of these simple wave functions, we have illustrated three general properties of orbital angular momentum in quantum physics:

- Orbital angular momentum in quantum physics is quantized and the natural unit for angular momentum is

$$\hbar = 1.055 \times 10^{-34} \text{ J s.}$$

- The orbital angular momentum of a quantum particle is at best a fuzzy vector. We have only been able to specify precisely the magnitude and just one of the components of orbital angular momentum. This is because the components of angular momentum are non-compatible observables as discussed generally in Chapter 7.

- A quantum particle with specific orbital angular momentum properties has a wave function with a specific angular shape. If the orbital angular momentum is zero the wave function is spherically symmetric, and if the orbital angular momentum is non-zero the wave function has angular dependence.

Spherical harmonics

So far we have considered wave functions to be functions of the Cartesian coordinates x, y and z. In practice, it is more useful to consider wave functions to be functions of the spherical polar coordinates r, θ and ϕ illustrated Fig. 8.5. This figure shows that the Cartesian and spherical coordinates of the point P are related by

$$x = r \sin \theta \cos \phi, \quad y = r \sin \theta \sin \phi, \quad \text{and} \quad z = r \cos \theta.$$

When a quantum state is represented by a wave function $\Psi(r, \theta, \phi)$, the dependence on θ and ϕ specifies an angular shape that determines the orbital angular momentum properties of the state. In fact, all possible orbital angular momentum properties can be described using simultaneous eigenfunctions of $\hat{L}^2$ and $\hat{L}_z$. These eigenfunctions are called *spherical harmonics*. They are denoted $Y_{l,m_l}(\theta, \phi)$ and they satisfy the eigenvalue equations:

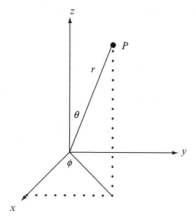

Fig. 8.5   The spherical polar coordinates (r, θ, ϕ) of the point P.

$$\hat{L}^2 Y_{l,m_l} = l(l+1)\hbar^2 Y_{l,m_l} \quad \text{and} \quad \hat{L}_z Y_{l,m_l} = m_l \hbar Y_{l,m_l}, \tag{8.23}$$

where the quantum numbers l and m_l can take on the values $l = 0, 1, 2, \ldots$ and $m_l = -l, \ldots, +l$.

These eigenfunctions are orthogonal because they satisfy

$$\int Y^*_{l',m'_l} Y_{l,m_l} \, d\Omega = 0 \quad \text{if} \quad l' \neq l \text{ and } m'_l \neq m_l \tag{8.24}$$

and they are usually normalized so that

$$\int |Y_{l,m_l}|^2 \, d\Omega = 1. \tag{8.25}$$

In these integrals $d\Omega$ is the solid angle

$$d\Omega = \sin\theta \, d\theta \, d\phi$$

and the limits of integration are from $\theta = 0$ to $\theta = \pi$ and from $\phi = 0$ to $\phi = 2\pi$.

Explicit forms of the spherical harmonics with $l = 0, l = 1$, and $l = 2$ are given in Table 8.1. If we compare these with the wave functions given by Eqs. (8.18) and (8.19), we see that

$$\psi_{(0,0)} \propto Y_{0,0}(\theta, \phi)$$

and that

$$\psi_{(1,0)} \propto Y_{1,0}(\theta, \phi) \quad \text{and} \quad \psi_{(1,\pm1)} \propto Y_{1,\pm1}(\theta, \phi).$$

TABLE 8.1   Spherical harmonics with $l = 0, 1$ and 2

Spherical harmonics as functions of θ and ϕ	Spherical harmonics as functions of x, y and z
$Y_{0,0} = \sqrt{\dfrac{1}{4\pi}}$	$Y_{0,0} = \sqrt{\dfrac{1}{4\pi}}$
$Y_{1,0} = \sqrt{\dfrac{3}{4\pi}} \cos\theta$	$Y_{1,0} = \sqrt{\dfrac{3}{4\pi}} \dfrac{z}{r}$
$Y_{1,\pm1} = \mp\sqrt{\dfrac{3}{8\pi}} \sin\theta \, e^{\pm i\phi}$	$Y_{1,\pm1} = \mp\sqrt{\dfrac{3}{8\pi}} \dfrac{x \pm iy}{r}$
$Y_{2,0} = \sqrt{\dfrac{5}{16\pi}}(3\cos^2\theta - 1)$	$Y_{2,0} = \sqrt{\dfrac{5}{16\pi}} \dfrac{3z^2 - r^2}{r^2}$
$Y_{2,\pm1} = \mp\sqrt{\dfrac{15}{8\pi}} \sin\theta \cos\theta \, e^{\pm i\phi}$	$Y_{2,\pm1} = \mp\sqrt{\dfrac{15}{8\pi}} \dfrac{(x \pm iy)z}{r^2}$
$Y_{2,\pm2} = \sqrt{\dfrac{15}{32\pi}} \sin^2\theta \, e^{\pm 2i\phi}$	$Y_{2,\pm2} = \sqrt{\dfrac{15}{32\pi}} \dfrac{x^2 - y^2 \pm 2ixy}{r^2}$

We also note that spherical harmonics have a simple dependence on the azimuthal angle ϕ, given by

$$Y_{l,m_l}(\theta, \phi) = F_{l,m_l}(\theta)\, e^{im_l\phi}, \tag{8.26}$$

but that the θ dependence becomes increasingly complicated as l increases.

The angular shape of the position probability density for a particle with angular momentum quantum numbers l and m_l is given by $|Y_{l,m_l}(\theta, \phi)|^2$. The angular shapes for $l = 0$ and $l = 1$ were shown in Fig. 8.4 and the more complex shapes for $l = 2$ and $l = 3$ are shown in Figs. 8.6 and 8.7. We note that there is no dependence on the azimuthal angle ϕ, but the dependence on the angle θ becomes more complex as l increases.

Linear superposition

We have already emphasized that each orbital angular momentum eigenfunction has a specific angular shape. We shall now describe how these shapes form a complete set of angular shapes. To illustrate this idea in the simplest possible context, we shall focus exclusively, for the moment, on the ϕ dependence of the wave function and suppress any reference to the r and θ coordinates.

Any complex function $\psi(\phi)$ in the interval $0 \leq \phi \leq 2\pi$ can be expressed as the Fourier series

$$\psi(\phi) = \sum_n c_n\, e^{in\phi}$$

where n is an integer that runs from $-\infty$ to $+\infty$, and where the coefficients c_n are given by

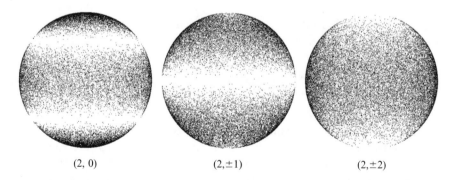

(2, 0)   (2, ±1)   (2, ±2)

Fig. 8.6   The position probability densities on the surface of a sphere for a particle with quantum numbers (l, m_l) equal to $(2, 0)$, $(2, \pm 1)$ and $(2, \pm 2)$. (This figure was produced with the permission of Thomas D. York.)

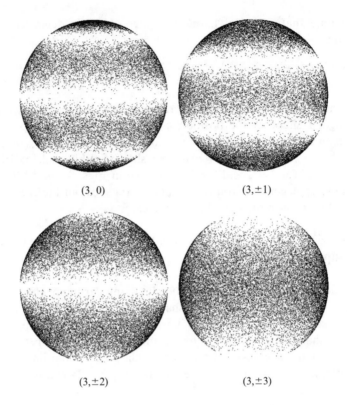

(3, 0)   (3, ±1)

(3, ±2)   (3, ±3)

Fig. 8.7   The position probability densities on the surface of a sphere for a particle with quantum numbers (l, m_l) equal to $(3, 0)$, $(3, \pm 1)$, $(3, \pm 2)$, and $(3, \pm 3)$. (This figure was produced with the permission of Thomas D. York.)

$$c_n = \frac{1}{2\pi} \int_0^{2\pi} \mathrm{e}^{-in\phi} \psi(\phi)\, \mathrm{d}\phi.$$

To bring the notation into line with the conventions of quantum physics, we shall rewrite this Fourier series as

$$\psi(\phi) = \sum_{m_l} c_{m_l} Z_{m_l}(\phi), \quad \text{where} \quad Z_{m_l}(\phi) = \frac{\mathrm{e}^{im_l\phi}}{\sqrt{2\pi}}, \tag{8.27}$$

where m_l is an integer that runs from $-\infty$ to $+\infty$.

In problem 5, we shall show that the basis functions $Z_{m_l}(\phi)$ are eigenfunctions of $\hat{L}_z$ with eigenvalues $m_l\hbar$. Thus Eq. (8.27) is yet another example of the principle of linear superposition in quantum mechanics, which states that any quantum state is a linear superposition of other quantum states; in this case, a linear superposition of quantum states with definite values for L_z. The coefficients c_{m_l} are probability amplitudes for L_z, because $|c_{m_l}|^2$ is the probability that the measured value of the z component of orbital angular momentum is equal to $m_l\hbar$.

In a similar way, the θ and ϕ dependence of any wave function can be expressed as a generalized Fourier series involving basis functions which are eigenfunctions of $\hat{L}^2$ and $\hat{L}_z$. These eigenfunctions form a complete set of three-dimensional angular shapes so that any wave function $\psi(r, \theta, \phi)$ can be expressed as

$$\psi(r, \theta, \phi) = \sum_{l=0}^{l=\infty} \sum_{m_l=-l}^{m=+l} c_{l,m_l}(r) Y_{l,m_l}(\theta, \phi). \tag{8.28}$$

By using the orthogonality and normalization conditions for spherical harmonics, Eqs. (8.24) and (8.25), we can show that the coefficients $c_{l,m_l}(r)$ of this series are given by

$$c_{l,m_l}(r) = \int Y^*_{l,m_l}(\theta, \phi)\psi(r, \theta, \phi) \, d\Omega. \tag{8.29}$$

These coefficients are probability amplitudes for orbital angular momentum; in fact, the probability that the particle is found between r and $r + dr$ with orbital angular momentum $L = \sqrt{l(l+1)}\hbar$ and $L_z = m_l\hbar$ is given by $|c_{l,m_l}(r)|^2 r^2 \, dr$.

As an example, let us consider the wave function $\psi'_{(1,0)}$ given by Eq. (8.22). By using Table 8.1, we find

$$\psi'_{(1,0)} = -\sqrt{\frac{2\pi}{3}} \frac{R(r)}{r} [Y_{1,+1}(\theta, \phi) + Y_{1,-1}(\theta, \phi)].$$

Because this is a linear superposition of spherical harmonics with $l = 1, m_l = +1$ and $l = 1, m_l = -1$, a measurement of the magnitude and z component of the orbital angular momentum can have two possible outcomes: $L = \sqrt{2}\hbar, L_z = +\hbar$ or $L = \sqrt{2}\hbar, L_z = -\hbar$. Because the magnitudes of the coefficients of the superposition are the same, each of these outcomes has the same probability.

The linear superposition given by Eq. (8.28) provides a useful representation of the wave function of a scattered particle. In this case, the function $c_{l,m_l}(r)$ is called a *partial wave*. It can be decomposed into an incoming spherical wave and an outgoing spherical wave, and the effect of scattering is to cause a shift in the phase of the outgoing spherical wave. The analogous phase shift in a one-dimensional scattering process was considered in Section 5.1, but in a three-dimensional scattering process, there is a phase shift for each orbital angular momentum. These phase shifts can be used to calculate the scattering cross-section.

PROBLEMS 8

1. A particle has orbital angular momentum given by the quantum number $l = 3$ and spin angular momentum given by the quantum number $s = 1$.

   (a) How many distinct states are there with different values for the z components of the orbital and spin angular momenta?

   (b) What are the possible values for the quantum number j that describes the total angular momentum of the particle?

   (c) How many distinct states are there with different values for the magnitude and z component of total angular momentum?

   (Note that the rules for the addition of angular momenta given by Eq. (8.6) are such that, when angular momenta with quantum numbers l and s are combined to give total angular momenta with quantum numbers $j = l+s, l+s-1, \ldots |l-s|$, then the number of distinct states with different values for m_l and m_s is equal to the number of distinct states with different values for j and m_j.)

2. A classical electron moves in a circle of radius 1 mm with velocity 1 mm s^{-1}.

   (a) What is the value of the quantum number l which gives a quantized angular momentum close to the angular momentum of this classical electron?

   (b) How many discrete values are possible for the z component of this orbital angular momentum?

   (c) How closely spaced are these values as a fraction of the magnitude of the orbital angular momentum?

3. The ground state of the hydrogen atom consists of an electron and a proton with zero orbital angular momentum and with magnetic moments given by Eq. (8.8) and Eq. (8.12). The atom is placed in a magnetic field of 0.5 T.

   (a) Explain why, if the effect of the proton magnetic moment can be ignored, the ground state energy is split into two energy levels. What is the spacing between these energy levels in eV?

   (b) Explain why each of these magnetic energy levels consists of two closely spaced energy levels if the effect of the proton magnetic moment is taken into account, and if the external magnetic field is large compared with any internal field. What is the spacing between these closely spaced levels in eV?

4. Two particles of mass m are attached to the ends of a massless rod of length a. The system is free to rotate in three dimensions about its centre of mass.

(a) Write down an expression for the classical kinetic energy of rotation of the system, and show that the quantum rotational energy levels are given by

$$E_l = \frac{l(l+1)\hbar^2}{ma^2} \quad \text{with} \quad l = 0, 1, 2, \dots.$$

(b) What is the degeneracy of the lth energy level?

(c) The H_2 molecule consists of two protons separated by a distance of 0.075 nm. Find the energy needed to excite the first excited rotational state of the molecule.

5. (a) By considering the relation between Cartesian and spherical polar co-ordinates,

$$x = r \sin \theta \cos \phi, \quad y = r \sin \theta \sin \phi, \quad \text{and} \quad z = r \cos \theta,$$

and the chain rule

$$\frac{\partial \psi}{\partial \phi} = \frac{\partial \psi}{\partial x} \frac{\partial x}{\partial \phi} + \frac{\partial \psi}{\partial y} \frac{\partial y}{\partial \phi} + \frac{\partial \psi}{\partial z} \frac{\partial z}{\partial \phi},$$

show that the operator for the z component of the orbital angular momentum of a particle,

$$\hat{L}_z = -i\hbar \left(x \frac{\partial}{\partial y} - y \frac{\partial}{\partial x} \right),$$

can be rewritten as

$$\hat{L}_z = -i\hbar \frac{\partial}{\partial \phi}.$$

(b) Verify that

$$Z_{m_l}(\phi) = \frac{e^{im_l\phi}}{\sqrt{2\pi}}$$

is an eigenfunction of $\hat{L}_z$ with eigenvalue $m_l\hbar$.

(c) Explain why it is not unreasonable to assume that any wave function satisfies the condition

$$\psi(r, \theta, \phi) = \psi(r, \theta, \phi + 2\pi).$$

Show that this condition implies that m_l is an integer.

(d) Show that the integral

$$\int_0^{2\pi} Z_{m'_l}^*(\phi)Z_{m_l}(\phi)\,\mathrm{d}\phi$$

is equal to one if $m'_l = m_l$ and zero if $m'_l \neq m_l$.

(e) Show that, if

$$\psi(r, \theta, \phi) = \sum_{m_l} c_{m_l}(r, \theta)Z_{m_l}(\phi),$$

then

$$c_{m_l}(r, \theta) = \int_0^{2\pi} Z_{m_l}^*(\phi)\psi(r, \theta, \phi)\,\mathrm{d}\phi.$$

6. Consider a wave function with azimuthal dependence

$$\psi(r, \theta, \phi) \propto \sin 2\phi \cos \phi.$$

What are the possible outcomes of a measurement of the z component of the orbital angular momentum and what are the probabilities of these outcomes?

(This question can be tackled using the expression for c_{m_l} given in the last part of the preceding problem, but the simplest approach is to rewrite $\sin 2\phi$ and $\cos \phi$ in terms of complex exponentials and tidy up.)

7. Consider a wave function with azimuthal dependence

$$\psi(r, \theta, \phi) \propto \cos^2 \phi.$$

What are the possible outcomes of a measurement of the z component of the orbital angular momentum and what are the probabilities of these outcomes?

8. In spherical polar coordinates the operator for the square of the orbital angular momentum is

$$\hat{L}^2 = -\hbar^2 \left(\frac{\partial^2}{\partial\theta^2} + \frac{\cos\theta}{\sin\theta}\frac{\partial}{\partial\theta} + \frac{1}{\sin^2\theta}\frac{\partial^2}{\partial\phi^2} \right).$$

Show that the simultaneous eigenfunctions of $\hat{L}^2$ and $\hat{L}_z$ have the form

$$Y_{l,m_l}(\theta,\phi) = F_{l,m_l}(\theta)\, e^{im_l\phi},$$

where the function $F_{l,m_l}(\theta)$ satisfies the differential equation

$$\frac{d^2 F_{l,m_l}}{d\theta^2} + \frac{\cos\theta}{\sin\theta}\frac{d F_{l,m_l}}{d\theta} + \left(l(l+1) - \frac{m_l^2}{\sin^2\theta} \right) F_{l,m_l} = 0.$$

It can be shown that finite solutions of this differential equation, in the range $0 \leq \theta \leq \pi$, only exist if the quantum numbers l and m_l take on the values given by $l = 0, 1, 2, ..$ and $m_l = -l, \ldots, +l$. Find, by substitution, the values of l and m_l for which the following functions are solutions:

$$F_a(\theta) = A, \quad F_b(\theta) = B\cos\theta \quad \text{and} \quad F_c(\theta) = C\sin\theta,$$

where A, B and C are constants.

9. Reconsider the energy eigenfunctions for a three-dimensional harmonic oscillator given in Section 6.5. Note that, by using Table 8.1, it is possible to form linear combinations of these eigenfunctions to give simultaneous eigenfunctions of energy, L^2 and L_z.

(a) Verify that the eigenfunction with energy $\frac{3}{2}\hbar\omega$ has orbital angular momentum quantum numbers $l = 0$ and $m_l = 0$.

(b) Construct eigenfunctions with energy $\frac{5}{2}\hbar\omega$ with quantum numbers $l = 1$ and $m_l = -1, 0 + 1$.

(c) Construct eigenfunctions with energy $\frac{7}{2}\hbar\omega$ with quantum numbers $l = 2$ and $m_l = -2, -1, 0, +1, +2$, and one eigenfunction with energy $\frac{7}{2}\hbar\omega$ with orbital quantum numbers $l = 0$ and $m_l = 0$.

9

The hydrogen atom

Just as the solar system provided the first meaningful test of the laws of classical mechanics, the hydrogen atom provided the first meaningful test of the laws of quantum mechanics. The hydrogen atom is the simplest atom, consisting of an electron with charge $-e$ and a nucleus with charge $+e$. To a first approximation, the nucleus, with a mass much larger than the electron mass, can be taken as a fixed object. This means that it should be possible to understand the properties of the hydrogen atom by solving a one-particle quantum mechanical problem, the problem of an electron in the Coulomb potential energy field

$$V(r) = -\frac{e^2}{4\pi\epsilon_0 r}.\tag{9.1}$$

In this chapter we shall find the energy eigenvalues and eigenfunctions for such an electron and use these results to describe the main features of the hydrogen atom. We will then show how this description can be improved by including small effects due to relativity and the motion of the nucleus.

9.1 CENTRAL POTENTIALS

We shall begin by considering the general problem of a particle in a central potential which is a potential, like the Coulomb potential, that only depends on the distance of the particle from a fixed origin.

Classical mechanics of a particle in a central potential

Let us consider a classical particle of mass m with vector position $\mathbf{r}$, momentum $\mathbf{p}$ and orbital angular momentum $\mathbf{L} = \mathbf{r} \times \mathbf{p}$ with respect to a fixed origin. If the particle moves in a central potential $V(r)$, it is subject to a force which is given by

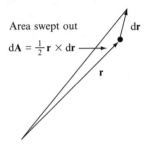

Area swept out

$dA = \frac{1}{2}\mathbf{r} \times d\mathbf{r}$

Fig. 9.1   The radius vector $\mathbf{r}$ of a particle with angular momentum $\mathbf{L} = m\mathbf{r} \times d\mathbf{r}/dt$ sweeps out a vector area $d\mathbf{A} = (L/2m)\,dt$ in time dt.

$$\mathbf{F} = -\frac{dV}{dr}\mathbf{e}_r$$

where $\mathbf{e}_r$ is a unit vector in the direction of $\mathbf{r}$. Because this force acts along the radius vector $\mathbf{r}$, the torque acting on the particle, $\mathbf{N} = \mathbf{r} \times \mathbf{F}$, is zero and the particle moves with constant angular momentum $\mathbf{L}$.

The geometrical implications of constant angular momentum can be understood by considering Fig. 9.1 which shows that the vector area swept by the radius vector $\mathbf{r}$ in time dt is given by

$$d\mathbf{A} = \frac{\mathbf{L}}{2m}\,dt.$$

This implies that, when the angular momentum $\mathbf{L}$ is a constant vector, the particle moves in a fixed plane with a radius vector which sweeps out area at a constant rate of $L/2m$.

The momentum $\mathbf{p}$ of a particle moving in a plane has two independent components which may be conveniently taken to be the radial and transverse components

$$p_r = m\frac{dr}{dt} \quad \text{and} \quad p_t = \frac{L}{m}.$$

By writing the kinetic energy $\mathbf{p}^2/2m$ on terms of p_r and p_t, we find that the constant total energy of the particle is given by

$$E = \frac{p_r^2}{2m} + \frac{L^2}{2mr^2} + V(r). \tag{9.2}$$

We note that the energy of the particle can be viewed as the sum of two terms, the radial kinetic energy $p_r^2/2m$ and an effective potential energy of the form

$$V_e(r) = \frac{L^2}{2mr^2} + V(r). \tag{9.3}$$

The effective force corresponding to this effective potential acts in the radial direction and has magnitude

$$F_e = -\frac{dV_e}{dr} = \frac{L^2}{mr^3} - \frac{dV}{dr}.$$

The term L^2/mr^3, which equals mv^2/r for a particle moving with speed v in circle of radius r with angular momentum $L = mrv$, represents an outward centrifugal force. Thus, the term $L^2/2mr^2$ in the effective potential (9.3) can be thought of as either a centrifugal potential energy or as a transverse kinetic energy.

The most important example of classical motion in a central potential is planetary motion. A planet of mass m moves around the sun with gravitational potential energy

$$V(r) = -\frac{GmM_\odot}{r}$$

where $M_\odot$ is the solar mass and G is Newton's fundamental constant of gravity. Books on classical mechanics show that planets in the solar system move in elliptic orbits. If a planet could shed excess energy its orbit would eventually become circular. If it could acquire energy so that it could just escape from the sun, its orbit would become a parabola, and when the energy is higher still its orbit would be a hyperbola. Circles, ellipses, parabolas and hyperbolas are *conic sections* which are best studied by taking a ripe, conical pear and slicing it up.

Planetary motion provided the first extra-terrestrial test of the laws of classical mechanics. These classical laws passed the test with flying colours because the orbital angular momentum of a planet is many orders of magnitude greater than the fundamental quantum unit of angular momentum $\hbar$; for example, the orbital angular momentum of planet earth is a stupendous $3 \times 10^{74}\ \hbar$.

The hydrogen atom provides another example of motion in a central potential. In this case, an electron with charge $-e$ moves around a nucleus with charge e in the Coulomb potential Eq. (9.1). When the electron has an angular momentum much greater than $\hbar$, classical mechanics can be used and the electron traces an orbit which is a conic section. But, when the angular momentum is comparable with $\hbar$, quantum mechanics must be used and the electron is described by a quantum state with uncertain properties.

Quantum mechanics of a particle in a central potential

The quantum states of a particle in a central potential are described by a wave function $\Psi(r, \theta, \phi, t)$. We shall focus on quantum states with sharply defined energy E which, according to Section 4.3, have wave functions of the form

$$\Psi(r, \theta, \phi, t) = \psi(r, \theta, \phi) \, e^{-iEt/\hbar}, \tag{9.4}$$

where $\psi(r, \theta, \phi)$ is an energy eigenfuction satisfying the eigenvalue equation

$$\left[-\frac{\hbar^2}{2m} \nabla^2 + V(r) \right] \psi = E\psi. \tag{9.5}$$

This partial differential equation in three independent variables r, θ and ϕ may be greatly simplified if we assume that the quantum state has, in addition to definite energy E, definite angular momentum properties of the type described in Chapter 8. In particular, if we assume that the magnitude of the orbital angular momentum is $L = \sqrt{l(l+1)}\hbar$ and its z-component is $L_z = m_l\hbar$, where l and m_l are quantum numbers which could take on the values $l = 0, 1, 2 \ldots$ and $m_l = -l, \ldots, l$, the eigenfunctions have the form

$$\psi(r, \theta, \phi) = R(r) Y_{l, m_l}(\theta, \phi). \tag{9.6}$$

In this equation, $Y_{l, m_l}(\theta, \phi)$ is a simultaneous eigenfunction of $\hat{L}^2$ and $\hat{L}_z$ satisfying Eq. (8.23) and $R(r)$ is an unknown function of r. If we substitute Eq. (9.6) into Eq. (9.5), use the identities

$$\nabla^2 \psi = \frac{1}{r} \frac{\partial^2(r\psi)}{\partial r^2} + \frac{1}{r^2} \left[\frac{\partial^2\psi}{\partial\theta^2} + \frac{\cos\theta}{\sin\theta} \frac{\partial\psi}{\partial\theta} + \frac{1}{\sin^2\theta} \frac{\partial^2\psi}{\partial\phi^2} \right]$$

and

$$\hat{L}^2 = -\hbar^2 \left(\frac{\partial^2}{\partial\theta^2} + \frac{\cos\theta}{\sin\theta} \frac{\partial}{\partial\theta} + \frac{1}{\sin^2\theta} \frac{\partial^2}{\partial\phi^2} \right),$$

and also use Eq. (8.23), we obtain the following ordinary differential equation for $R(r)$:

$$-\frac{\hbar^2}{2mr} \frac{d^2(rR)}{dr^2} + \left[\frac{l(l+1)\hbar^2}{2mr^2} + V(r) \right] R = ER. \tag{9.7}$$

By introducing a radial function $u(r)$, defined by

$$R(r) \equiv \frac{u(r)}{r},$$ (9.8)

we obtain

$$-\frac{\hbar^2}{2m}\frac{d^2 u}{dr^2} + \left[\frac{l(l+1)\hbar^2}{2mr^2} + V(r)\right]u = Eu.$$ (9.9)

This important equation is called the *radial Schrödinger equation*. It describes a particle with angular momentum $L = \sqrt{l(l+1)}\hbar$ which behaves like a particle in a one-dimensional effective potential of the form

$$V_e(r) = \frac{l(l+1)\hbar^2}{2mr^2} + V(r).$$ (9.10)

If we compare this potential with the analogous effective potential in classical mechanics given in Eq. (9.3), we see that the first term, $l(l+1)\hbar^2/2mr^2$, can either be thought of as a kinetic energy associated with transverse motion or as a centrifugal potential that arises from the orbital angular momentum of the particle.

When solutions of the radial Schrödinger equation (9.9) are sought, the boundary condition

$$u(r) = 0 \quad \text{at} \quad r = 0$$

must be imposed to ensure that the function $R(r) = u(r)/r$, and hence the actual three-dimensional eigenfunction given by Eq. (9.6), is finite at the origin. In addition, bound state solutions, which describe a particle that cannot escape to infinity, must also satisfy the boundary condition

$$u(r) \to 0 \quad \text{as} \quad r \to \infty.$$

Bound states only exist if the effective potential $V_e(r)$, Eq. (9.10), is sufficiently attractive. We shall label these states by a quantum number $n_r = 0, 1, 2, \ldots$ which will be shown to be equal to the number nodes of the radial eigenfunction $u(r)$ between $r = 0$ and $r = \infty$. This means a bound state of a particle in a central potential can always be specified by three quantum numbers n_r, l and m_l and that the eigenfunction has the form

$$\psi_{n_r, l, m_l}(r, \theta, \phi) = \frac{u_{n_r, l}(r)}{r} \, Y_{l, m_l}(\theta, \phi).$$ (9.11)

By using the normalization condition (8.25) for spherical harmonics, we can easily show that the eigenfunction $\psi_{n_r, l, m_l}(r, \theta, \phi)$ is normalized if the radial eigenfunction $u_{n_r, l}(r)$ obeys the condition

$$\int_0^\infty |u_{n_r, l}|^2 \, dr = 1. \tag{9.12}$$

The energy of these bound states will be denoted $E_{n_r, l}$. By thinking of this energy as the sum of three terms, the average radial kinetic energy, the average transverse kinetic energy and the average Coulomb energy, we can see that states with higher l and higher n_r have higher energies. The energy $E_{n_r, l}$ increases with l because the average transverse kinetic energy is given by

$$\int_0^\infty u_{n_r, l}^*(r) \left(\frac{l(l+1)\hbar^2}{2mr^2} \right) u_{n_r, l}(r) \, dr;$$

and the energy $E_{n_r, l}$ also increases with n_r because the average radial kinetic energy, which is given by

$$\int_0^\infty u_{n_r, l}^*(r) \left(-\frac{\hbar^2}{2m} \frac{d^2}{dr^2} \right) u_{n_r, l}(r) \, dr,$$

increases as the number of radial nodes increases.[1]

Before considering the explicit expressions for the energy levels and eigenfunctions of a particle in a Coulomb potential, we pause and reconsider how we obtained the radial Schrödinger equation Eq. (9.9). The crucial step was to seek a quantum state with definite E, L and L_z for a particle in a central potential. Such states must exist because this step has had a successful outcome: it has led to Eq. (9.9) which can be solved to give sensible energy eigenvalues and eigenfunctions. Readers who have studied compatible observables in Chapter 7 should not be surprised by this success. In general, E, L and L_z can be taken as three compatible observables which uniquely define a quantum state of a particle in a central potential.

In addition, a quantum state of a particle with definite energy in a central potential has another observable property with a definite value. It is called *parity*. Eigenfunctions with the property

$$\psi(-\mathbf{r}) = +\psi(\mathbf{r}), \tag{9.13}$$

---

[1] We encountered a similar behaviour when we considered the energy levels of one-dimensional potential wells in Chapters 4, 5 and 6. In general, more nodes in a wave function means higher energy in quantum mechanics, just as more nodes in a classical normal mode means higher frequency.

are said to have even parity, and eigenfunctions with the property

$$\psi(-\mathbf{r}) = -\psi(\mathbf{r}) \tag{9.14}$$

are said to have odd parity. By using Table 8.1, we can easily show that an eigenfunction with definite orbital angular momentum given by Eq. (9.6), i.e. by

$$\psi(r, \theta, \phi) = R(r) Y_{l, m_l}(\theta, \phi),$$

has even parity when $l = 0$ and $l = 2$ and odd parity when $l = 1$. It can be shown that, in general, the parity of a particle in a central potential is even when l is even and odd when l is odd.

Parity is one of the simplest observables in quantum mechanics, but, because it is an observable with no classical analogue, it is often perceived as mysterious. Energy and parity are compatible observables whenever the Hamiltonian is unchanged when the coordinates undergo a reflection through the origin. Because this is true for all Hamiltonians which do not involve the weak nuclear interaction, parity has an important role in classifying quantum states in atomic, nuclear and particle physics.

9.2   QUANTUM MECHANICS OF THE HYDROGEN ATOM

In this section we shall give a brief description of the hydrogen atom using the quantum mechanical concepts introduced in the last section.

Because the hydrogen atom is essentially an electron in a Coulomb potential and the Coulomb potential is a central potential, bound states of atomic hydrogen may be taken to have definite orbital angular momentum properties given by $L = \sqrt{l(l+1)}\hbar$ and $L_z = m_l \hbar$. These states have wave functions of the form

$$\Psi_{n_r, l, m_l}(r, \theta, \phi, t) = \psi_{n_r, l, m_l}(r, \theta, \phi) \exp(-iE_{n_r, l}t/\hbar)$$

with

$$\psi_{n_r, l, m_l}(r, \theta, \phi) = \frac{u_{n_r, l}(r)}{r} Y_{l, m_l}(\theta, \phi),$$

where $u_{n_r, l}(r)$ is an eigenfunction given by the radial Schrödinger equation (9.9) for an electron in a Coulomb potential

$$V(r) = -\frac{e^2}{4\pi\epsilon_0 r}.$$

Specifically, the radial eigenfunction $u_{n_r,l}(r)$ is a solution of the differential equation

$$\frac{\hbar^2}{2m_e}\frac{d^2u_{n_r,l}}{dr^2} + \left[\frac{l(l+1)\hbar^2}{2m_er^2} - \frac{e^2}{4\pi\epsilon_0 r}\right]u_{n_r,l} = E_{n_r,l}u_{n_r,l}, \qquad (9.15)$$

which satisfies the boundary conditions

$$u_{n_r,l}(r) = 0 \quad \text{at} \quad r = 0 \quad \text{and at} \quad r = \infty. \qquad (9.16)$$

The qualitative features of the energy levels given by the eigenvalue problem defined by Eqs. (9.15) and (9.16) may be deduced by considering the effective potential that occurs in Eq. (9.15),

$$V_e(r) = \frac{l(l+1)\hbar^2}{2m_er^2} - \frac{e^2}{4\pi\epsilon_0 r}. \qquad (9.17)$$

The shape of this potential for electrons with different values for the orbital angular momentum quantum number l are shown in Fig. 9.2. We see that, for non-zero values of l, the effective potential is attractive at large r and repulsive at small r. By setting dV_e/dr to zero, we can easily show that $V_e(r)$ has a minimum value of

$$V_e(r) = -\frac{E_R}{l(l+1)} \quad \text{at} \quad r = l(l+1)a_0, \qquad (9.18)$$

where a_0 and E_R are the natural units of length and energy in atomic physics, defined as follows: the Bohr radius

$$a_0 = \left[\frac{4\pi\epsilon_0}{e^2}\right]\frac{\hbar^2}{m_e} = 0.529 \times 10^{-10} \text{ m} \qquad (9.19)$$

and the Rydberg energy

$$E_R = \frac{e^2}{8\pi\epsilon_0 a_0} = 13.6 \text{ eV}. \qquad (9.20)$$

The minimum given by Eq. (9.18) implies that bound states with angular momentum $L = \sqrt{l(l+1)}\hbar$ have energies somewhere between $E = -E_R/l(l+1)$ and $E = 0$. It also implies that the spatial extent of bound state eigenfunctions with low angular momentum is of the order of a_0 and that the eigenfunctions extend to larger distances when the angular momentum increases. When the angular momentum greatly exceeds $\hbar$, we expect many

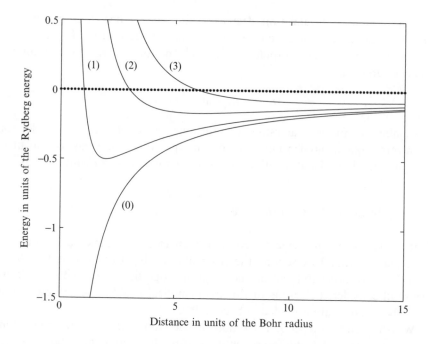

Fig. 9.2   The effective potential energy $V_e(r)$ for an electron in a hydrogen atom with orbital angular momentum quantum numbers $l = 0$, 1, 2 and 3. The unit of distance is the Bohr radius a_0, defined in Eq. (9.19), and the unit of energy is the Rydberg energy E_R, defined in Eq. (9.20). The effective potential for an electron with quantum number l has a minimum value of $-E_R/l(l+1)$ at $r = l(l+1)a_0$ which, in classical physics, corresponds to the energy and radius of a circular orbit of an electron with angular momentum $L = \sqrt{l(l+1)}\hbar$.

bound states with closely spaced energy levels that correspond to the circular and elliptic orbits of classical mechanics.

We also see from Fig. 9.2 that the effective potential is purely attractive for an electron with zero angular momentum. In classical mechanics, such an electron simply plunges towards the proton and there are no stable bound states. But there are stable bound states with zero angular momentum in quantum mechanics, whose existence can be understood by using the uncertainty principle.

The uncertainty principle implies that an electron localized in a region of size r has an uncertain momentum of the order of $\hbar/r$ and an average kinetic energy which is at least of the order of $\hbar^2/2m_e r^2$. This means that the least energy of an electron with zero orbital angular momentum in a region of size r near a proton is roughly given by

$$E \approx \frac{\hbar^2}{2m_e r^2} - \frac{e^2}{4\pi\epsilon_0 r}.$$

As the region of localization decreases, this energy decreases because the potential energy decreases, but eventually the kinetic energy of localization, $\hbar^2/2m_e r^2$, increases more rapidly. As a result the total energy has a minimum value of about

$$E \approx -E_R \quad \text{at} \quad r \approx a_0.$$

This minimum provides an estimate of the lowest possible energy of an electron with zero angular momentum in a Coulomb potential and suggests that there are quantum bound states with energies in the range $E \approx -E_R$ and $E = 0$.

Energy levels and eigenfunctions

The energy levels and eigenfunctions for an electron bound in a Coulomb potential are found by solving the eigenvalue problem defined by Eqs.(9.15) and (9.16). In order to focus on the physical properties of the hydrogen atom, we shall consider the results of this mathematical problem before describing how these results are obtained in Section 9.7.

We shall show in Section 9.7 that an electron with angular momentum $L = \sqrt{l(l+1)}\hbar$ in a Coulomb potential has an infinite number of bound states with energies given by

$$E_{n_r,l} = -\frac{E_R}{(n_r + l + 1)^2} \quad \text{with} \quad n_r = 0, 1, 2, 3, \ldots \tag{9.21}$$

The quantum number n_r is called the *radial quantum number*. These energy levels are illustrated in Fig. 9.3.

As expected, there are bound states with zero and non-zero angular momentum. Also the energy levels are very closely spaced if the angular momentum is very large, indicating a correspondence with a continuum of classical bound-state energies.

Unexpectedly, many of the energy levels in Fig. 9.3, those with the same value for $n_r + l$, have the same energy. Because of this degeneracy, the energy levels of the hydrogen atom are usually given as

$$E_n = -\frac{E_R}{n^2} \tag{9.22}$$

where E_R is the Rydberg energy and n is a quantum number defined by

$$n = n_r + l + 1. \tag{9.23}$$

This quantum number is called the *principal quantum number* and can take on the values $n = 1, 2, 3, \ldots$

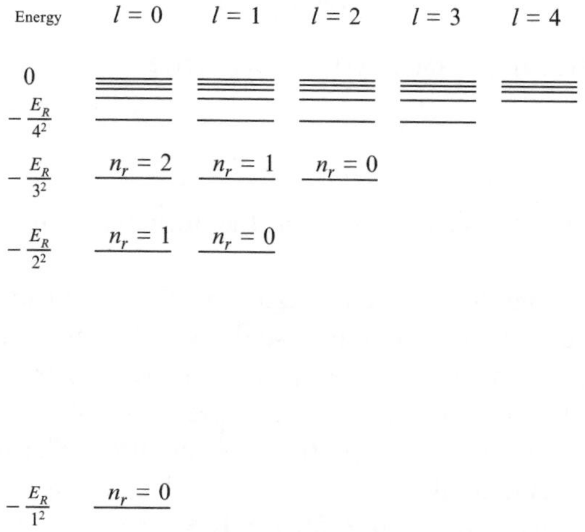

Fig. 9.3 The energy levels given by Eq. (9.21) for the bound states of an electron in a hydrogen atom. For each value of the orbital angular momentum quantum number $l = 0, 1, 2,$, there is an infinite stack of energies $E_{n_r, l}$ with radial quantum number $n_r = 0, 1, 2, \ldots$. Note that, because bound states with the same value for $n_r + l$ have the same energy, these energy levels are often given as $E_n = -E_R/n^2$ where E_R is the Rydberg energy and n is defined by $n = n_r + l + 1$; the quantum number n is called the *principal quantum number* and can take on the values $n = 1, 2, 3, \ldots$. There is also a continuum of unbound or ionized states with positive energy.

We shall also show in Section 9.7 that the radial eigenfunction $u_{n_r, l}(r)$ belonging to the energy eigenvalue $E_{n_r, l}$ has three characteristics:

(1) In Section 5.1 we found that the eigenfunction of a particle in a one-dimensional square well with binding energy $\epsilon = \hbar^2 \alpha^2 / 2m$ falls off exponentially like $e^{-\alpha x}$. In a similar way, the radial eigenfunction of an electron in a Coulomb potential with binding energy

$$\epsilon = \frac{E_R}{n^2} = \frac{\hbar^2}{2m_e} \frac{1}{n^2 a_0^2}$$

falls off exponentially at large r like

$$u_{n_r, l}(r) \propto e^{-r/na_0}.$$

(2) Because of the singular nature at $r = 0$ of the centrifugal potential $l(l+1)\hbar^2/2m_e r^2$, the behaviour of the eigenfunction at small r is governed by the orbital angular momentum and is given by

$$u_{n_r, l}(r) \propto r^{l+1}.$$

(3) Finally, because the radial quantum number n_r denotes the number of nodes between $r = 0$ and $r = \infty$, the eigenfunction $u_{n_r, l}(r)$ is proportional to a polynomial with n_r zeros. If this polynomial is denoted by $p_{n_r, l}(r)$, we have

$$u_{n_r, l}(r) \propto p_{n_r, l}(r).$$

By combining these three characteristics, we arrive at a radial eigenfunction of the form

$$u_{n_r, l}(r) = N p_{n_r, l}(r) r^{l+1} \, e^{-r/na_0}, \tag{9.24}$$

where N is a constant which ensures that the normalization condition (9.12) is satisfied.

Explicit expressions for the radial eigenfunctions with low values for the angular momentum quantum number l and low values for the radial quantum number n_r are given in Table 9.1, and some of these eigenfunctions are illustrated in Fig. 9.4. To conform with the conventions of atomic physics these eigenfunctions are labelled using spectroscopic notation. This notation employs the principal quantum number $n = n_r + l + 1$ and a letter to designate the value of l; the letter s is used for $l = 0$, p for $l = 1$, d for $l = 2$, and f for $l = 3$. The historical origin of this notation dates back to the early days of atomic physics when spectral lines were labelled s for sharp, p for principal, d for diffuse and f for fundamental.

TABLE 9.1   Normalized radial eigenfunctions for low-lying states of the hydrogen atom

Spectroscopic notation	Radial eigenfunction $u_{n_r, l}(r)$
1s	$u_{0,0}(r) = \dfrac{2}{\sqrt{a_0}} \left(\dfrac{r}{a_0} \right) e^{-r/a_0}$
2s	$u_{1,0}(r) = \dfrac{1}{\sqrt{2a_0}} \left[1 - \dfrac{1}{2}\left(\dfrac{r}{a_0} \right) \right] \left(\dfrac{r}{a_0} \right) e^{-r/2a_0}$
3s	$u_{2,0}(r) = \dfrac{2}{3\sqrt{3a_0}} \left[1 - \dfrac{2}{3}\left(\dfrac{r}{a_0} \right) + \dfrac{2}{27}\left(\dfrac{r}{a_0} \right)^2 \right] \left(\dfrac{r}{a_0} \right) e^{-r/3a_0}$
4s	$u_{3,0}(r) = \dfrac{1}{4\sqrt{a_0}} \left[1 - \dfrac{3}{4}\left(\dfrac{r}{a_0} \right) + \dfrac{1}{8}\left(\dfrac{r}{a_0} \right)^2 - \dfrac{1}{192}\left(\dfrac{r}{a_0} \right)^3 \right] \left(\dfrac{r}{a_0} \right) e^{-r/4a_0}$
2p	$u_{0,1}(r) = \dfrac{1}{2\sqrt{6a_0}} \left(\dfrac{r}{a_0} \right)^2 e^{-r/2a_0}$
3p	$u_{1,1}(r) = \dfrac{8}{27\sqrt{6a_0}} \left[1 - \dfrac{1}{6}\left(\dfrac{r}{a_0} \right) \right] \left(\dfrac{r}{a_0} \right)^2 e^{-r/3a_0}$
4p	$u_{2,1}(r) = \dfrac{1}{16} \sqrt{\dfrac{5}{3a_0}} \left[1 - \dfrac{1}{4}\left(\dfrac{r}{a_0} \right) + \dfrac{1}{80}\left(\dfrac{r}{a_0} \right)^2 \right] \left(\dfrac{r}{a_0} \right)^2 e^{-r/4a_0}$
3d	$u_{0,2}(r) = \dfrac{4}{81\sqrt{30a_0}} \left(\dfrac{r}{a_0} \right)^3 e^{-r/3a_0}$
4d	$u_{1,2}(r) = \dfrac{1}{64\sqrt{5a_0}} \left[1 - \dfrac{1}{12}\left(\dfrac{r}{a_0} \right) \right] \left(\dfrac{r}{a_0} \right)^3 e^{-r/4a_0}$

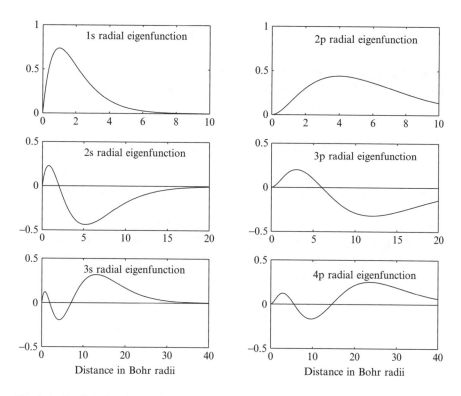

Fig. 9.4 Radial eigenfunctions $u_{n_r, l}(r)$ for an electron in the hydrogen atom with radial quantum numbers $n_r = 0$, 1, and 2 and with angular momentum quantum numbers $l = 0$ and 1. The eigenfunctions have been labelled using spectroscopic notation ns and np, where n is the principal quantum number $n = n_r + 1 + l$ and s denotes $l = 0$ and p denotes $l = 1$. Note that the unit of distance is the Bohr radius a_0 and that different scales are used for eigenfunctions with different values of n_r.

9.3   SIZES AND SHAPES

The size and shape of a quantum state of atomic hydrogen can be determined by considering the most probable locations of the electron in the atom. For a state with eigenfunction $\psi_{n_r, l, m_l}(r, \theta, \phi)$, the probability of finding the electron with coordinates (r, θ, ϕ) in a volume element $d^3\mathbf{r}$ is

$$|\psi_{n_r, l, m_l}(r, \theta, \phi)|^2 \, d^3\mathbf{r}.$$

We can easily find the radial probability distribution for the electron. To do so, we use $d^3\mathbf{r} = r^2 \, dr \, d\Omega$, where $d\Omega = \sin\theta \, d\theta \, d\phi$ is an element of solid angle, and express the eigenfunction as

$$\psi_{n_r, l, m_l}(r, \theta, \phi) = \frac{u_{n_r, l}(r)}{r} Y_{l, m_l}(\theta, \phi),$$

where $Y_{l, m_l}(\theta, \phi)$ is a spherical harmonic which obeys the normalization condition Eq. (8.25). The probability of finding the electron at a distance between r and $r + dr$ from the nucleus is then given by

$$\left| \frac{u_{n_r, l}(r)}{r} \right|^2 r^2 \, dr \int |Y_{l, m_l}(\theta, \phi)|^2 \, d\Omega = |u_{n_r, l}(r)|^2 \, dr.$$

Thus, the radial shape of the quantum state is described by a radial probability density $|u_{n_r, l}(r)|^2$.

Radial probability densities for states with $n_r = 0, 1$ and 2 and with $l = 0$ and 1 are shown in Fig. 9.5. We note that the radial extent increases as n_r increases and as l increases. This may be confirmed by considering the mean radius of a state with quantum numbers n_r and l which is given by

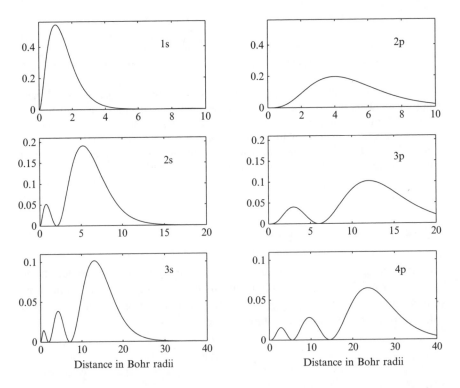

Fig. 9.5 Radial probability densities for the 1s, 2s, 3s, 2p, 3p and 4p states of the hydrogen atom. Note that the unit of distance is the Bohr radius a_0 and that a different scale is used for states with a different number of radial nodes.

$$\langle r \rangle_{n_r,l} = \int_0^\infty r |u_{n_r,l}(r)|^2 \, dr.$$

This integral may be evaluated by using the mathematical properties of the eigenfunctions $u_{n_r,l}(r)$ to give an expression,

$$\langle r \rangle_{n_r,l} = \tfrac{1}{2}[3n_r^2 + 6n_r(l+1) + (l+1)(2l+3)]a_0, \qquad (9.25)$$

which explicitly shows how the radial extent increases with n_r and l.

By using explicit forms for $u_{n_r,l}(r)$ and $Y_{l,m_l}(\theta,\phi)$ we can evaluate the probability density $|\psi_{n_r,l,m_l}(r,\theta,\phi)|^2$ and explore both the radial and angular shapes of a hydrogen atom state with quantum numbers n_r, l, m_l. First, we note that the angular shape does not depend on the azimuthal angle ϕ because, as indicated by Eq. (8.26), the spherical harmonic $Y_{l,m_l}(\theta,\phi)$ has the ϕ dependence $e^{im_l\phi}$. Thus, the probability density of the state does not change if it is rotated about the z axis. This means that the size and shape of the state may be fully specified by showing the most likely positions of the electron on any vertical plane that passes through the z axis, as demonstrated in Figs. 9.6 and 9.7.

In Figs. 9.6 and 9.7 we show the sizes and shapes of the 3p and 3d states of atomic hydrogen. The 3p states in Fig. 9.6 have one radial node and an angular dependence given by $|Y_{1,m_l}(\theta,\phi)|^2$ with $m_l = 1, 0$ and -1. The 3d states in Fig. 9.7 have no radial nodes and the angular shape of a state with $l = 2$.

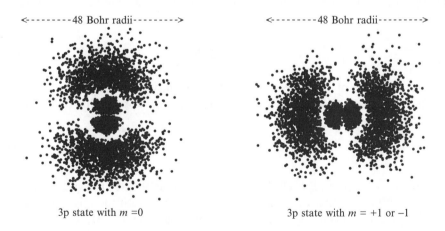

<----------48 Bohr radii--------->   <----------48 Bohr radii--------->

3p state with $m = 0$ 3p state with $m = +1$ or -1

Fig. 9.6 The size and shape of the 3p states of the hydrogen atom with a z component of orbital angular momentum equal to $m\hbar$. These states have rotational symmetry about the z axis. The density of dots is proportional to the probability of finding an electron on a vertical plane passing through the z axis. These pictures were produced by selecting a point on the plane at random and deciding to plot or not to plot in accordance with the value of $|\psi_{n_r,l,m_l}(r,\theta,\phi)|^2$ at that point.

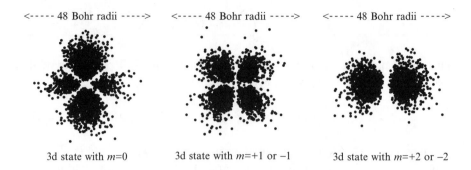

Fig. 9.7   The size and shape of the 3d states of the hydrogen atom with a z component of orbital angular momentum equal to $m\hbar$.

9.4   RADIATIVE TRANSITIONS

When a hydrogen atom interacts with an electromagnetic field, quantum states with quantum numbers n, l and m_l are, in general, no longer stationary states with definite energy, and radiative transitions between these states may take place in which electromagnetic energy is either absorbed or emitted.

The most probable radiative transitions are called *electric dipole transitions*. They are caused by an interaction of the electric field component $\mathbf{E}$ of the electromagnetic field with the operator describing the electric dipole moment of the electron–nucleus system. The electric dipole operator is $\mathbf{d} = -e\mathbf{r}$, where $\mathbf{r}$ is the vector position operator for the electron in the atom, and the interaction is given by

$$\hat{H}_I = -\mathbf{d} \cdot \mathbf{E}. \tag{9.26}$$

In the presence of this interaction, the probability for a transition between states with quantum numbers n_i, l_i, m_{l_i} and n_f, l_f, m_{l_f} is proportional to

$$\left| \int \psi^*_{n_f, l_f, m_{l_f}}(\mathbf{r}) \, \hat{H}_I \, \psi_{n_i, l_i, m_{l_i}}(\mathbf{r}) \, \mathrm{d}^3 \mathbf{r} \right|^2. \tag{9.27}$$

We can easily prove that electric dipole transitions always involve a change in parity by showing that the integral in Eq. (9.27) is zero if the initial and final states have the same parity. We show this by considering the effect of changing the integration variable from $\mathbf{r}$ to $-\mathbf{r}$. The interaction $\hat{H}_I = -\mathbf{d} \cdot \mathbf{E}$ changes sign, but the sign of the eigenfunction, $\psi_{n_i, l_i, m_{l_i}}(\mathbf{r})$ or $\psi_{n_f, l_f, m_{l_f}}(\mathbf{r})$, is unchanged if the eigenfunction has even parity and it is changed if the eigenfunction has odd parity, as shown by Eqs. (9.13) and (9.14). Thus, when both eigenfunctions have the same parity, the integrand in Eq. (9.27) changes sign when the integration variable $\mathbf{r}$ is changed to $-\mathbf{r}$ and this implies that the integral must be zero.

It can also be shown, by noting that the angular dependence of the eigenfunctions $\psi_{n_f, l_f, m_{l_f}}(\mathbf{r})$ and $\psi_{n_i, l_i, m_{l_i}}(\mathbf{r})$ are given by spherical harmonics, that the integral in Eq. (9.27), and hence the probability of transition, is zero unless the difference $\Delta l = l_f - l_i$ is $+1$ or -1. This means that all electric dipole transitions in the hydrogen atom also obey the *selection rule*

$$\Delta l = \pm 1. \tag{9.28}$$

The electric dipole transitions between low-lying states of the hydrogen atom are shown as dotted lines in Fig. 9.8, where spectroscopic notation, 1s, 2s, 2p, etc. has been used to label the levels corresponding to states with different values for the principal quantum number n and orbital angular momentum quantum number l; for example, 2s corresponds to $n = 2$ and $l = 0$ and 2p corresponds to $n = 2$ and $l = 1$.

The transitions shown in Fig. 9.8 may be *induced* or *spontaneous*. Induced transitions between states with energy E_{n_i} and E_{n_f} occur strongly when the atom interacts with an external electromagnetic field which oscillates with an angular frequency ω which satisfies the resonant condition

$$\hbar\omega = |E_{n_f} - E_{n_i}|;$$

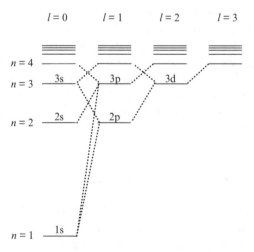

Fig. 9.8   Electric dipole radiative transitions between low-lying energy levels of the hydrogen atom with different values for the quantum numbers n and l. Spectroscopic notation, 1s, 2s, 2p, etc. has been used to label the energy levels; for example, 2s corresponds to $n = 2$ and $l = 0$ and 2p corresponds to $n = 2$ and $l = 1$. We note that the electric dipole transitions shown by the dotted lines obey the $\Delta l = \pm 1$ selection rule given in Eq. (9.28).

electromagnetic energy is absorbed when $E_{n_f} > E_{n_i}$ and it is emitted when $E_{n_f} < E_{n_i}$. Spontaneous transitions are seemingly not caused by anything, but they are really caused by the interaction of the atom with a *quantized radiation field* which is always present, even when the atom is entirely isolated. As a result of this interaction, an atom with energy E_{n_i} decays to an atom with energy E_{n_f} by emitting a photon with energy $\epsilon = E_{n_i} - E_{n_f}$. Such transitions give rise to spectral lines with wavelengths λ given by

$$\frac{hc}{\lambda} = E_R\left(\frac{1}{n_f^2} - \frac{1}{n_i^2}\right). \tag{9.29}$$

As mentioned in Section 1.3, these lines form a series of lines in the ultra-violet called the Lyman series, with wavelengths given by

$$\frac{hc}{\lambda} = E_R\left(\frac{1}{1^2} - \frac{1}{n_i^2}\right), \quad \text{with} \quad n_i = 2, 3, 4\ldots,$$

a series of lines in the visible called the Balmer series, with wavelengths given by

$$\frac{hc}{\lambda} = E_R\left(\frac{1}{2^2} - \frac{1}{n_i^2}\right), \quad \text{with} \quad n_i = 3, 4, 5\ldots,$$

and other series with longer wavelengths.

Inspection of Fig. 9.8 shows that the 2s state of the hydrogen atom cannot decay by electric dipole radiation because a 2s $\rightarrow$ 1s transition would violate the $\Delta l = \pm 1$ rule given in Eq. (9.28). In fact, the 2s state is a *metastable* stable state with a long lifetime which eventually decays to the 1s state by a mechanism which is less probable than an electric dipole transition. Whereas the mean time for a spontaneous 2p $\rightarrow$ 1s transition is 1.6×10^{-9} seconds, the mean time for a spontaneous 2s $\rightarrow$ 1s transition is a lengthy 0.14 seconds.

9.5   THE REDUCED MASS EFFECT

So far we have assumed that the mass of the nucleus of the hydrogen atom is sufficiently large so that its motion can be ignored. In fact, this motion gives rise to small but important effects which can be incorporated by introducing the concept of the *reduced mass of the electron–nucleus system*.[2]

---

[2] We encountered the reduced mass of the proton-proton system at the end of Section 5.2 and the reduced mass of a diatomic molecule in Section 6.4.

To understand the reduced mass effect in atomic hydrogen, we consider the classical energy of an electron of mass m_e and a nucleus of mass m_N interacting via a Coulomb potential

$$E = \frac{p_e^2}{2m_e} + \frac{p_N^2}{2m_N} - \frac{e^2}{4\pi\epsilon_0 r},$$

where p_e and p_N are the magnitudes of momenta of the electron and the nucleus, and r is the distance between the electron and the nucleus. In the centre-of-mass frame we can set $p_e = p_N = p$ and obtain an expression for the energy of the form

$$E = \frac{p^2}{2\mu} - \frac{e^2}{4\pi\epsilon_0 r}, \quad \text{where} \quad \mu = \frac{m_e m_N}{m_e + m_N}.$$

This expression shows that a classical electron–nucleus system in the centre of mass frame acts like a single particle of mass μ. This mass is called the reduced mass of the electron–nucleus system.

In a similar way a quantum electron–nucleus system also acts like a single particle with reduced mass μ. This means that we can take account of the motion of the nucleus in the hydrogen atom by simply replacing the electron mass m_e by the reduced mass μ. In particular, the length and energy scales given by the Bohr radius a_0, Eq. (9.19), and the Rydberg energy E_R, Eq. (9.20), are modified slightly and become

$$a_0' = \left[\frac{4\pi\epsilon_0}{e^2}\right] \frac{\hbar^2}{\mu} = \frac{m_e}{\mu} a_0$$

and

$$E_R' = \frac{e^2}{8\pi\epsilon_0 a_0'} = \frac{\mu}{m_e} E_R,$$

and all the quantitative results given in Eqs. (9.21), (9.22), (9.24) and (9.25), in Table 9.1, and in Figs. 9.3, 9.4 and 9.5 are modified accordingly. Most importantly, the modified energy levels of the hydrogen atom are given by

$$E_n' = -\frac{E_R'}{n^2} = \frac{\mu}{m_e} \frac{E_R}{n^2}. \tag{9.30}$$

The reduced mass effect in the hydrogen atom is apparent when the spectral lines of ordinary hydrogen are compared with the spectral lines of heavy hydrogen. For an atom of ordinary hydrogen, the nucleus is a proton with mass $1836m_e$ and the reduced mass is $(1836/1837)m_e$, whereas for an atom of heavy hydrogen the nucleus is a deuteron with mass $3671m_e$ and the reduced mass is $(3671/3672)m_e$. The tiny difference between these two reduced masses gives rise to an observable difference in the wavelengths of the spectral lines emitted by these atoms. For example, the H_α line of the Balmer series, which arises from a transition from a state with $n = 3$ to one with $n = 2$, has a wavelength λ given by

$$\frac{hc}{\lambda} = \frac{\mu}{m_e} E_R \left[\frac{1}{2^2} - \frac{1}{3^2} \right].$$

For ordinary hydrogen this gives $\lambda = 656.4686$ nm and, for heavy hydrogen, it gives $\lambda = 656.2899$ nm. In fact, Harold Urey discovered heavy hydrogen in 1934 in an experiment which revealed that each line in the Balmer series was accompanied by a faint line due to the small admixture of heavy hydrogen that is present in naturally occurring hydrogen.

9.6   RELATIVISTIC EFFECTS

In this section we shall confirm that, to a good approximation, the electron in a hydrogen atom is non-relativistic, but that relativistic effects give rise to small and significant corrections to the energy levels.

We can confirm that the electron in the hydrogen atom is approximately non-relativistic by estimating its average momentum. Because the uncertainty in the position of the electron is of the order of the Bohr radius a_0, the uncertainty in its momentum, and hence its average momentum, is of the order of $p_0 = \hbar/a_0$. Using Eq. (9.19) we find that

$$p_0 = \alpha m_e c, \tag{9.31}$$

where α is a dimensionless constant, called the *fine structure constant*, given by

$$\alpha = \frac{e^2}{4\pi\epsilon_0 \hbar c} = \frac{1}{137.035\,989\,5}. \tag{9.32}$$

The small numerical value of the fine structure constant has an important implication for atomic physics. It implies that atomic electrons, like the electron in a hydrogen atom, have momenta which are small compared with $m_e c$ and that non-relativistic physics is a good approximation.

We can estimate the magnitude of relativistic corrections to the hydrogen atom by considering the relation between the relativistic energy ϵ and momentum p of an electron,

$$\epsilon = \sqrt{m_e^2 c^4 + p^2 c^2}.$$

If $p \ll m_e$, we can use the binomial theorem to give the approximate expression

$$\epsilon = m_e c^2 + \frac{p^2}{2m_e} - \frac{1}{8}\left(\frac{p}{m_e c}\right)^4 m_e c^2.$$

The first term is the rest-mass energy, the second is the non-relativistic kinetic energy and the third term is the leading relativistic correction to the kinetic energy. We note that, because the average momentum of an electron in a hydrogen atom is of the order of $p_0 = \alpha m_e c$, the relativistic correction to its kinetic energy is of the order of

$$\langle E_{rel}\rangle \approx -\frac{1}{8}\left(\frac{p_0}{m_e c}\right)^4 m_e c^2 = -\frac{1}{8}\alpha^4 m_e c^2. \tag{9.33}$$

Corrections of a similar magnitude also arise from the interaction of the spin magnetic moment of the electron with a magnetic field caused by the relative motion of the nucleus and the electron. This interaction is called the *spin–orbit interaction*.

The magnitude of the spin–orbit interaction can be estimated by considering a classical electron moving around the nucleus in a circle of radius r. If the electron has velocity v and orbital angular momentum $L = m_e r v$, it will take time

$$\tau = \frac{2\pi r}{v} = \frac{2\pi m_e r^2}{L}$$

to complete one orbit. Just as we perceive the sun to move around the earth, the electron perceives a nucleus which appears to move in a circle of radius r with period τ. Because the nucleus has charge e, an electric current $I = e/\tau$ circles the electron constituting a current loop. Using the standard formula for the magnetic field at the centre of a circular current loop of radius r, $B = \mu_0 I/2r$, and the relation $c^2 = 1/\sqrt{\epsilon_0 \mu_0}$, we find that the electron perceives a magnetic field of magnitude

$$B = \frac{e}{4\pi\epsilon_0 m_e c^2 r^3} L. \tag{9.34}$$

Because the electron has a spin magnetic moment, which according to Eq. (8.8) is given by

$$\boldsymbol{\mu} = -2\frac{e}{2m_e}\mathbf{S},$$

there is an energy of interaction given by

$$E_{mag} = -\boldsymbol{\mu} \cdot \mathbf{B} = \frac{e^2}{4\pi\epsilon_0 m_e^2 c^2 r^3} \mathbf{L} \cdot \mathbf{S}.$$

This estimate has given the right order of magnitude for the spin–orbit interaction. A more careful calculation, which takes account of the acceleration of the electron, introduces a factor of one-half and gives

$$E_{mag} = \frac{e^2}{8\pi\epsilon_0 m_e^2 c^2 r^3} \mathbf{L} \cdot \mathbf{S}. \tag{9.35}$$

The quantitative effects of the spin–orbit interaction on the energy levels of the hydrogen atom can be obtained by evaluating expectation values of E_{mag}. However, two general aspects of the spin–orbit interaction are worth mentioning here:

(1) The spin–orbit interaction is really a relativistic effect and we can confirm this by showing that it leads to a correction which is comparable with the relativistic correction to the kinetic energy given by Eq. (9.33)

$$\langle E_{rel} \rangle \approx -\frac{1}{8} \alpha^4 m_e c^2.$$

Using the fine structure constant α defined in Eq. (9.32), we rewrite Eq. (9.35) for the spin–orbit energy as

$$E_{mag} = \alpha \frac{\hbar}{2m_e^2 c} \frac{\mathbf{L} \cdot \mathbf{S}}{r^3}.$$

For a low-lying state of the hydrogen atom, we can assume that the expectation value of $\mathbf{L} \cdot \mathbf{S}/r^3$ is of the order of $\hbar^2/a_0^3$. Rewriting the Bohr radius a_0 defined in Eq. (9.19) in terms of the fine structure constant as

$$a_0 = \frac{1}{\alpha} \frac{\hbar}{m_e c},$$

we find that the spin–orbit energy of a low-lying hydrogen atom state is of the order of

$$\langle E_{mag} \rangle \approx \alpha^4 m_e c^2.$$

(2) The spin–orbit interaction implies that quantum states of the hydrogen atom with definite energy have definite values for J^2, L^2 and S^2, where

$J = L + S$ is the total angular momentum due to the orbital and spin angular momenta of the electron. We can confirm this by using

$$\mathbf{J} \cdot \mathbf{J} = (\mathbf{L} + \mathbf{S}) \cdot (\mathbf{L} + \mathbf{S}) = L^2 + S^2 + 2\mathbf{L} \cdot \mathbf{S}$$

to show that the spin–orbit interaction is proportional to

$$\mathbf{L} \cdot \mathbf{S} = \frac{J^2 - L^2 - S^2}{2}.$$

As a quantitative illustration of the combined effect in the hydrogen atom of the relativistic correction to the kinetic energy and of the relativistic spin–orbit interaction, we consider quantum states with principal quantum number $n = 2$. Because of the spin–orbit interaction, these states will have definite energy if they have definite magnitudes for spin, orbital and total angular momenta given by

$$S = \sqrt{s(s+1)}\hbar, \quad L = \sqrt{l(l+1)}\hbar \quad \text{and} \quad J = \sqrt{j(j+1)}\hbar,$$

where l, s and j are quantum numbers. The rule for the addition of spin and orbital angular momenta, which is given by Eq. (8.6), implies that, in general, j can take the values

$$j = l + s, \ l + s - 1, \dots, \ |l - s|.$$

For hydrogen atom states with $n = 2$, the orbital quantum number l can be 0 or 1. The $n = 2$ states with $l = 0$ and $s = \frac{1}{2}$ have $j = \frac{1}{2}$ and are denoted by $2s_{1/2}$, and the $n = 2$ states with $l = 1$ and $s = \frac{1}{2}$ have $j = \frac{1}{2}$ or $j = \frac{3}{2}$ and are denoted by $2p_{1/2}$ and $2p_{3/2}$.

If the spin–orbit interaction were absent, all states of the hydrogen atom with $n = 2$ would have the same energy

$$E_2 = -\frac{E_R}{2^2} = -\frac{\alpha^2}{8} m_e c^2,$$

where E_R is the Rydberg energy and α is the fine structure constant. When relativistic and spin–orbit corrections are calculated using a theory called perturbation theory, these states are found to have energies that depend on the quantum numbers l, s and j. These energies are given by

$$E(2s_{1/2}) = E_2 - \tfrac{5}{64} \alpha^4 m_e c^2,$$
$$E(2p_{1/2}) = E_2 - \tfrac{5}{64} \alpha^4 m_e c^2,$$

and

$$E(2p_{3/2}) = E_2 - \tfrac{1}{64}\,\alpha^4 m_e c^2.$$

We note that the difference in energies of the $2p_{3/2}$ and $2p_{1/2}$ states can be verified by observing a small difference in the wavelengths of the radiation emitted by the transitions $2p_{3/2} \to 1s_{1/2}$ and $2p_{1/2} \to 1s_{1/2}$. We also note that the $2s_{1/2}$ and $2p_{1/2}$ states are predicted to have the same energy, but that this degeneracy is removed by a small effect called the *Lamb shift* which arises from the quantum field properties of the electromagnetic field.

9.7   THE COULOMB EIGENVALUE PROBLEM

In this section we shall find the energy levels and eigenfunctions of an electron in a Coulomb potential by solving the eigenvalue problem defined by the differential equation (9.15) and the boundary conditions (9.16). *This section may be omitted without significant loss of continuity.*

As a first step we shall tidy up Eq. (9.15) by setting

$$r = q a_0 \quad \text{and} \quad E = -\gamma^2 E_R, \tag{9.36}$$

so that q is a dimensionless measure of distance and γ^2 is a dimensionless measure of the binding energy. If we use the definitions for a_0 and E_R, Eqs. (9.19) and (9.20), we find that the radial eigenfunction $u(q)$, when expressed as a function q, satisfies the differential equation

$$\frac{d^2 u}{dq^2} + \left[\frac{2}{q} - \frac{l(l+1)}{q^2} \right] u = \gamma^2 u \tag{9.37}$$

and the boundary conditions

$$u(q) = 0 \quad \text{at} \quad q = 0 \quad \text{and} \quad \text{at} \quad q = \infty. \tag{9.38}$$

Our next step is to find the behaviour of $u(q)$ at large q and at small q. At large q the differential equation (9.37) becomes

$$\frac{d^2 u}{dq^2} = \gamma^2 u.$$

The general solution is

$$u(q) = A\,e^{-\gamma q} + B\,e^{+\gamma q},$$

where A and B are constants, but to satisfy the boundary condition $u(q) \to 0$ as $q \to \infty$ we set $B = 0$ to give

$$u(q) = A\,\mathrm{e}^{-\gamma q} \quad \text{at large } q. \tag{9.39}$$

At small q, Eq. (9.37) becomes

$$\frac{\mathrm{d}^2 u}{\mathrm{d}q^2} - \frac{l(l+1)u}{q^2} = 0.$$

This equation has solutions of the form q^{l+1} and q^{-l}, as can be shown by substitution, but to ensure that the boundary condition $u(0) = 0$ is satisfied we must have

$$u(q) \propto q^{l+1} \quad \text{at small} \quad q. \tag{9.40}$$

Our third step is to seek an eigenfunction of the form

$$u(q) = f(q)q^{l+1}\,\mathrm{e}^{-\gamma q}, \tag{9.41}$$

where the function $f(q)$ does not invalidate the small q and large q behaviour given by Eqs. (9.40) and (9.39). If we substitute this expression into Eq. (9.37) and use

$$\frac{\mathrm{d}u}{\mathrm{d}q} = [(l+1)q^l - \gamma q^{l+1}]\mathrm{e}^{-\gamma q}f + q^{l+1}\mathrm{e}^{-\gamma q}\frac{\mathrm{d}f}{\mathrm{d}q}$$

to give

$$\frac{\mathrm{d}^2 u}{\mathrm{d}q^2} = [l(l+1)q^{l-1} - 2\gamma(l+1)q^l + \gamma^2 q^{l+1}]\mathrm{e}^{-\gamma q}f$$

$$+ 2[(l+1)q^l - \gamma q^{l+1}]\mathrm{e}^{-\gamma q}\frac{\mathrm{d}f}{\mathrm{d}q} + q^{l+1}\mathrm{e}^{-\gamma q}\frac{\mathrm{d}^2 f}{\mathrm{d}q^2},$$

we find that the function $f(q)$ satisfies the differential equation

$$q\frac{\mathrm{d}^2 f}{\mathrm{d}q^2} + 2[(l+1) - \gamma q]\frac{\mathrm{d}f}{\mathrm{d}q} + 2[1 - \gamma(l+1)]f = 0. \tag{9.42}$$

Acceptable solutions of Eq. (9.42) only exist for special values of the energy parameter γ. The simplest is

$$f(q) = \text{constant}$$

which is clearly a solution if $\gamma = 1/(l+1)$. If we use Eq. (9.36) to express q in terms of r and γ in terms of E, we find that this solution gives eigenfunctions and energy eigenvalues with radial quantum number $n_r = 0$ of the form

$$u_{0,l}(r) = Nr^{l+1}e^{-r/(l+1)a_0} \quad \text{and} \quad E_{0,l} = -\frac{E_R}{(l+1)^2}.$$

To find the other eigenfunctions and eigenvalues, we shall express $f(q)$ as an infinite power series,

$$f(q) = \sum_{s=0,1,2...} a_s q^s, \tag{9.43}$$

and substitute into Eq. (9.42). After some tedious tidying we obtain

$$\sum_{s=0,1,2,...} \{[s(s+1) + 2(s+1)(l+1)]a_{s+1} - [2\gamma(s+l+1) - 2]a_s\}q^s = 0.$$

Thus, the power series (9.43) is a solution of Eq. (9.42) if the coefficients of the series satisfy the relation

$$a_{s+1} = \left[\frac{2\gamma(s+l+1) - 2}{s(s+1) + 2(s+1)(l+1)}\right] a_s. \tag{9.44}$$

We note that this relation implies that

$$a_{s+1} \rightarrow \left[\frac{2\gamma}{s}\right] a_s \quad \text{as} \quad s \rightarrow \infty.$$

By comparing this behaviour with the behaviour of the function

$$q e^{+2\gamma q} = q \sum_{s=0}^{\infty} \frac{(2\gamma q)^s}{s!},$$

we conclude that, at large q, the function $f(q)$ is proportional to $q e^{+2\gamma q}$ and that, at large q, the function $u(q) = f(q)q^{l+1}e^{-\gamma q}$ is proportional to $q^{l+2}e^{+\gamma q}$. Because this form for $u(q)$ does not satisfy the boundary condition (9.39), the infinite power series (9.43) is not an acceptable solution of Eq. (9.42).

However, the series does give rise to an acceptable solution when the series terminates at a finite power of q to give a polynomial in q. To find the condition for a polynomial solution of degree n_r, we consider the relation between the coefficients a_{n_r+1} and a_{n_r} given by Eq. (9.44),

$$a_{n_r+1} = \left[\frac{2\gamma(n_r+l+1) - 2}{n_r(n_r+1) + 2(n_r+1)(l+1)}\right] a_{n_r}.$$

This relation shows that a_{n_r+1} is zero and all subsequent coefficients are also zero, if the energy parameter γ takes on the special value

$$\gamma = \frac{1}{n_r + l + 1}.$$

Thus, when γ has this value, the power series terminates at $a_{n_r} q^{n_r}$ to give a polynomial solution of Eq. (9.42) which we will denote by $p_{n_r, l}(q)$.
This polynomial gives rise to a solution of Eq. (9.37) of the form

$$u(q) = p_{n_r, l}(q) q^{a+1} e^{-\gamma q}$$

that obeys the boundary conditions (9.38). If we express q in terms of r and γ in terms of E, we obtain the eigenfunctions and eigenvalues,

$$u_{n_r, l}(r) = N p_{n_r, l}(r) r^{l+1} e^{-r/(n_r + l + 1)a_0} \quad \text{and} \quad E_{n_r, l} = -\frac{E_R}{(n_r + l + 1)^2},$$

which formed the basis of our discussion of the properties of atomic hydrogen in Section 9.2.

PROBLEMS 9

1. Consider an electron in a Coulomb potential

$$V(r) = -\frac{e^2}{4\pi\epsilon_0 r}$$

with wave function

$$\psi(\mathbf{r}) = N e^{-r/a}$$

where a is a constant.

(a) What is the orbital angular momentum of the electron?

(b) Show that the expectation values of the potential energy and of the kinetic energy of the electron are given by

$$\langle V \rangle = -\frac{e^2}{4\pi\epsilon_0 a} \quad \text{and} \quad \langle T \rangle = \frac{\hbar^2}{2m_e a^2}.$$

(c) Show that the expectation value of the total energy of the electron is a minimum when a is equal to the Bohr radius a_0. Find the value of this minimum.

(The integral

$$\int_0^\infty r^k e^{-\alpha r} \, dr = \frac{k!}{\alpha^{k+1}}$$

is useful. It is also useful in problems 3, 4 and 10.)

2. Consider the effective potential $V_e(r)$ given by Eq. (9.3) for a classical particle in a Coulomb potential

$$V(r) = -\frac{e^2}{4\pi\epsilon_0 r}.$$

(a) Show that at

$$r = r_c = \frac{L^2}{m}\left[\frac{4\pi\epsilon_0}{e^2}\right]$$

$V_e(r)$ has a minimum value of

$$V_e(r_c) = -\frac{L^2}{2mr_c^2}.$$

Hence, show by inspection of Eq. (9.2), that a particle with angular momentum L has a minimum energy when it is in a circular orbit with radius r_c.

(b) Show that if the energy is given by

$$E = -\frac{e^2}{8\pi\epsilon_0 a}$$

where a is a constant, then the particle has an orbit in which the distance from the origin ranges from $a - \sqrt{a^2 - ar_c}$ to $a + \sqrt{a^2 - ar_c}$.

3. The radial eigenfunction for the quantum state of lowest energy for an electron with angular momentum $L = \sqrt{l(l+1)}\hbar$ in a Coulomb potential is

$$u_{0,l}(r) = Nr^{l+1} e^{-r/(l+1)a_0}$$

where N is a constant.

(a) Show that the eigenfunction satisfies the normalization condition Eq. (9.12) if

$$N^2 = \left[\frac{2}{(l+1)a_0}\right]^{2l+3} \frac{1}{(2l+2)!}.$$

(b) Show that the most probable radius of the state is

$$r_{\text{most probable}} = (l+1)^2 a_0.$$

(c) Show that the average radius of the state is

$$\langle r \rangle = \frac{(2l+3)(l+1)a_0}{2}.$$

(d) Show that the average of the square of the radius is

$$\langle r^2 \rangle = \frac{(2l+4)(2l+3)(l+1)^2 a_0^2}{4}$$

and that the uncertainty in the radius Δr becomes small compared with $\langle r \rangle$ as $l \to \infty$.

(e) Show that in the limit $l \to \infty$, the most probable radius and the average radius are both equal to r_c, the radius derived in problem 2 for a circular orbit of a classical particle with angular momentum L.

4. The eigenfunction for the ground state of the hydrogen atom has the form

$$\psi_1(r) = N_1 e^{-r/a_0}$$

where a_0 is the Bohr radius, N_1 is a constant and r is the radial distance of the electron from the nucleus.

(a) By normalizing the integral in accordance with

$$\int_0^\infty |\psi_1(r)|^2 \, 4\pi r^2 \, dr = 1,$$

find the constant N_1.

(b) Given that the eigenfunction of an excited state has the form

$$\psi_2(r) = N_2(1 + \lambda r)e^{-r/2a_0},$$

use the orthogonality condition for ψ_1 and ψ_2 to find the constant λ, and sketch the eigenfunction $\psi_2(r)$.

5. In this problem you are asked to verify that $g_n = n^2$ gives the number of independent hydrogen states with energy $E_n = -E_R/n^2$, i.e. g_n is the degeneracy of the level E_n. To do so you need to note that for a given value of n, the quantum number l can equal $n - 1, n - 2, \ldots 0$, and that for a given value of l, the quantum number m_l can have $2l + 1$ values between $+l$ and $-l$.

   (a) Verify that $g_n = n^2$ is valid when the principal quantum number takes on the values 1 and 2.

   (b) Verify that, if $g_n = n^2$ is valid when $n = k$, then it is also valid when $n = k + 1$.

   Note (a) and (b) imply that $g_n = n^2$ is valid for all the possible values of the principal quantum number.

6. Rewrite Eq. (9.25) for the mean radius of hydrogen atom states in terms of the principal quantum number n and orbital angular momentum quantum number l. Find the mean radii for the 1s, 2s, 2p, 3s, 3p and 3d states.

7. Use the virial theorem, derived in problem 8 at the end of Chapter 7, to find the expectation values of the kinetic and potential energies of an electron in a hydrogen atom state with principal quantum number n.

8. Use the three-dimensional equivalent of Eq. (3.20),

$$\tilde{\psi}(\mathbf{p}) = \frac{1}{(2\pi\hbar)^{3/2}} \int e^{-\mathbf{p}\cdot\mathbf{r}/\hbar}\, \psi(\mathbf{r})\, d^3r,$$

to show that the momentum probability amplitude for an electron in the ground state of the hydrogen atom with wave function

$$\psi(\mathbf{r}) = \frac{1}{\sqrt{\pi a_0^3}} e^{-r/a_0}$$

is given by

$$\tilde{\psi}(\mathbf{p}) = \frac{2\sqrt{2}}{\pi} \frac{p_0^{5/2}}{(p^2 + p_0^2)^2},$$

where $p_0 = \hbar/a_0$.

Hence show that the most probable magnitude of the momentum of the electron is $p_0/\sqrt{3}$ and that its mean value is $8p_0/3\pi$.

9. According to Eq. (9.27), the probability for an electric dipole transition between hydrogen states with quantum numbers n_i, l_i, m_{l_i} and n_f, l_f, m_{l_f} is proportional to the modulus squared of the integral

$$\int \psi^*_{n_f, l_f, m_{l_f}}(\mathbf{r}) \, \hat{H}_I \, \psi_{n_i, l_i, m_{l_i}}(\mathbf{r}) \, \mathrm{d}^3\mathbf{r}.$$

Show that this integral is zero for a 2s → 1s transition, but that it is not zero for a 2p → 1s transition.

10. Show that the expectation values of the spin–orbit energy given by Eq.(9.35) for a $2p_{3/2}$ and a $2p_{1/2}$ state of a hydrogen atom are

$$\langle E_{mag} \rangle_{2p_{3/2}} = \frac{1}{96} \alpha^4 m_e c^2 \quad \text{and} \quad \langle E_{mag} \rangle_{2p_{1/2}} = -\frac{2}{96} \alpha^4 m_e c^2.$$

(Hint: use the eigenfunctions given in Table 9.1 to find the expectation value of $1/r^3$ and note that

$$\langle (J^2 - L^2 - S^2) \rangle = j(j+1) - l(l+1) - s(s+1)$$

for a state with quantum numbers j, l and s.)

Find the energy difference of the $2p_{3/2}$ and $2p_{1/2}$ states and show that the difference in wavelengths for the transitions $2p_{3/2} \to 1s_{1/2}$ and $2p_{1/2} \to 1s_{1/2}$ is 5.4×10^{-4} nm.

11. Positronium is a bound state of an electron with charge $-e$ and its anti-particle, a positron with charge $+e$.

(a) Given that the mass of the positron is equal to the mass of the electron, what is the reduced mass of the electron–positron system?

(b) Write down an expression for the energy levels of positronium.

(c) Given that the mean radius of the ground state of the hydrogen atom is $\frac{3}{2}a_0$, what is the mean radius of the ground state of positronium?

12. A muonic hydrogen atom is a bound state of a muon with charge $-e$ and a proton with charge $+e$.

(a) Given that the mass of the muon is $207m_e$, what is the reduced mass of the muon–proton system?

(b) Write down an expression for the energy levels of muonic hydrogen.

(c) Given that the mean radius of the ground state of an ordinary hydrogen atom is $\frac{3}{2}a_0$, what is the mean radius of the ground state of a muonic hydrogen atom?

13. It is possible for the electron in a hydrogen atom to get very close to the nucleus. To explore this possibility, consider a radius R which is much smaller than a_0 and show that an electron in a 1s state has a probability of about $4(R/a_0)^3$ of being found between $r = 0$ and $r = R$.

   Given that the nucleus is a proton with a radius of about 2×10^{-15} m, make a rough estimate of the probability that a 1s electron is inside the proton. Make a similar estimate for a 1s muon in a muonic hydrogen atom, and explain why the energy levels of a muonic hydrogen atom are more sensitive to the size of the proton than the energy levels of an ordinary atom of hydrogen.

14. By modifying the formulae for the hydrogen atom given in this chapter, write down the energy E_n of an electron with principal quantum number n in the Coulomb potential due to a point charge Ze,

$$V(r) = -\frac{Ze^2}{4\pi\epsilon_0 r}.$$

By inspection of Table 9.1, write down the 1s, 2s, and 2p radial eigenfunctions for this electron.

   What are the ionization energies of the one-electron ions He^+ and Li^{2+}?

   Explain why relativistic corrections are more important for He^+ and Li^{2+} than they are for the hydrogen atom.

15. Before you tackle this problem, you should work through problem 8 at the end of Chapter 4.

   Consider the electron in an atom of the heavy isotope of hydrogen, tritium. The nucleus has charge e, and, apart from a small correction due to the reduced mass effect, the electron has energies and eigenfunctions that are identical to those of an ordinary hydrogen atom. However, the nucleus of the atom, a triton, is unstable and decays by beta-decay to form a nucleus of ^3He. When it does so, the electron in the tritium atom suddenly finds itself in a new Coulomb potential, the potential due a nucleus with charge $2e$.

Assume that the electron is initially in the ground state of the tritium atom and show that

$$P = \frac{128}{a_0^6} \left[\int_0^\infty r^2 \, e^{-3r/a_0} \, dr \right]^2$$

is the probability that the electron is, after the decay, in the ground state of the He^+ ion.

Evaluate the integral and verify that this probability is 0.702.

10

Identical particles

In classical physics we can, in principle, keep track of particles as they move and maintain a record of which one is which, even if they are identical. This is not possible in quantum physics. Quantum particles with uncertain position and momentum cannot be tracked, and, if they are identical, they are truly indistinguishable. The indistinguishability of identical particles is a fundamental principle in quantum physics which gives rise to phenomena which have no classical analogue. We shall show how this principle leads to the important concept of *exchange symmetry* and to a classification of quantum particles into two types called *bosons* and *fermions*.

This chapter, like Chapter 7, will deal with concepts that are more abstract than those encountered elsewhere in this book. Even though these concepts are fundamental, they need not be studied in detail in order to understand atoms which is the topic covered in Chapter 11.

10.1 EXCHANGE SYMMETRY

We shall primarily develop an understanding of identical particles by considering a system of two particles p and q. All the properties of such a system may be extracted from a two-particle wave function $\Psi(\mathbf{r}_p, \mathbf{r}_q, t)$. For example, the joint probability for finding particle p in a volume element $d^3\mathbf{a}$ at $\mathbf{r}_p = \mathbf{a}$ and particle q in volume element $d^3\mathbf{b}$ at $\mathbf{r}_q = \mathbf{b}$ is given by

$$|\Psi(\mathbf{a}, \mathbf{b}, t)|^2 \, d^3\mathbf{a} \, d^3\mathbf{b}. \tag{10.1}$$

When the particles are identical, the wave function must give rise to identical properties for each particle. In particular, the probability density for finding particle p at $\mathbf{a}$ and particle q at $\mathbf{b}$ must be the same as the probability density for finding particle q at $\mathbf{a}$ and particle p at $\mathbf{b}$. This requirement is met by a wave function which obeys the condition

$$|\Psi(\mathbf{a}, \mathbf{b}, t)|^2 = |\Psi(\mathbf{b}, \mathbf{a}, t)|^2. \tag{10.2}$$

When this condition for an acceptable wave function is satisfied, $|\Psi(\mathbf{a}, \mathbf{b}, t)|^2 \, d^3\mathbf{a} \, d^3\mathbf{b}$ is the probability that one or other of the particles is found at $\mathbf{a}$ and the other is found at $\mathbf{b}$.[1]

Condition (10.2) implies that the function $\Psi(\mathbf{a}, \mathbf{b}, t)$ differs from the function $\Psi(\mathbf{b}, \mathbf{a}, t)$ by at most a phase factor; i.e. we must have

$$\Psi(\mathbf{a}, \mathbf{b}, t) = e^{i\delta}\Psi(\mathbf{b}, \mathbf{a}, t). \tag{10.3}$$

However, the phase factor $e^{i\delta}$ can only take on two possible values. We can show this by noting that the exchange of $\mathbf{a}$ and $\mathbf{b}$ in Eq. (10.3) gives

$$\Psi(\mathbf{b}, \mathbf{a}, t) = e^{i\delta}\Psi(\mathbf{a}, \mathbf{b}, t)$$

which, when Eq. (10.3) is used again, gives

$$\Psi(\mathbf{b}, \mathbf{a}, t) = e^{i\delta} \, e^{i\delta} \, \Psi(\mathbf{b}, \mathbf{a}, t).$$

This equation shows that $e^{i\delta} e^{i\delta}$ must be equal to one and that the phase factor $e^{i\delta}$ has two possible values:

$$e^{i\delta} = 1 \quad \text{or} \quad e^{i\delta} = -1. \tag{10.4}$$

Thus, any wave function for two identical particles must either be a symmetric function

$$\Psi(\mathbf{a}, \mathbf{b}, t) = +\Psi(\mathbf{b}, \mathbf{a}, t) \tag{10.5}$$

or an antisymmetric function

$$\Psi(\mathbf{a}, \mathbf{b}, t) = -\Psi(\mathbf{b}, \mathbf{a}, t). \tag{10.6}$$

These functions are said to have *definite exchange symmetry*, a property which ensures that identical particles cannot be distinguished.

---

[1] Some readers may consider our treatment of identical quantum particles to be contrived, because we label the identical particles and then we only accept wave functions for which the labelling has no physical consequences. If so desired, the contrived logic of using labels which cannot be observed can be avoided, but to do so one must use quantum field theory.

10.2 PHYSICAL CONSEQUENCES

We have shown that identical quantum particles must be described by wave functions that are either symmetric or antisymmetric when the particles are exchanged. We shall explore the physical consequences of this exchange symmetry by considering the simple example of two identical particles of mass m in a one-dimensional harmonic oscillator potential. We shall assume that the particles have no direct interaction with each other so that their Hamiltonian operator is given by

$$\hat{H}(x_p, x_q) = -\frac{\hbar^2}{2m}\frac{\partial^2}{\partial x_p^2} - \frac{\hbar^2}{2m}\frac{\partial^2}{\partial x_q^2} + \frac{1}{2}m\omega^2 x_p^2 + \frac{1}{2}m\omega^2 x_q^2. \qquad (10.7)$$

In problem 2 at the end of this chapter, we show that each of the particles may occupy single-particle states represented by eigenfunctions ψ_n with energies given by $E_n = (n + \frac{1}{2})\hbar\omega$ with $n = 0, 1, 2 \ldots$.

We first consider two particles in the same single-particle state. If this state has quantum number n, the total energy of the two particles is

$$E = E_n + E_n = (2n + 1)\hbar\omega$$

and the two-particle wave function with this energy is

$$\Psi^{(S)}(x_p, x_q, t) = \psi_n(x_p)\psi_n(x_q)\, e^{-i(E_n + E_n)t/\hbar}, \qquad (10.8)$$

where ψ_n is an eigenfunction for a particle with energy E_n in a harmonic oscillator potential; explicit expressions for the eigenfunctions with low energy are given in Table 6.1. We have labelled the two-particle wave function with the superscript S to indicate that it is symmetric under particle exchange which means that it is an acceptable wave function for two identical particles. We note that it is impossible to construct an antisymmetric wave function if both particles are in the same single-particle state. This implies that two identical particles with an antisymmetric wave function cannot occupy the same single-particle state.

We now consider two particles in two different single-particle states. When the particles occupy states with quantum number n and n', their energy is

$$E = E_n + E_{n'} = (n + n' + 1)\hbar\omega.$$

The wave function for two distinguishable particles with this energy may be given by

$$\Psi_1^{(D)}(x_p, x_q, t) = \psi_n(x_p)\psi_{n'}(x_q)\, e^{-i(E_n + E_{n'})t/\hbar}, \qquad (10.9)$$

by

$$\Psi_2^{(D)}(x_p, x_q, t) = \psi_n(x_q)\psi_{n'}(x_p)\, e^{-i(E_n + E_{n'})t/\hbar}, \tag{10.10}$$

or by a linear combination of the form

$$\Psi^{(D)}(x_p, x_q, t) = c_1 \Psi_1^{(D)}(x_p, x_q, t) + c_2 \Psi_2^{(D)}(x_p, x_q, t), \tag{10.11}$$

where c_1 and c_2 are constants. The label D has been used to denote that these wave functions can be used to describe distinguishable particles. We note that Eq. (10.11) specifies a strange wave function in which both particles are associated with both single-particle states. Indeed, provided the wave functions Ψ_1^D and Ψ_2^D are normalized, $|c_1|^2$ gives the probability that particle p is in state n and particle q is in state n' and $|c_2|^2$ gives the probability that particle q is in state n and particle p is in state n'. Because quantum states with wave functions like (10.11) are called *entangled states*, we shall occasionally refer to states described by wave functions like (10.9) and (10.10) as *untangled states*. Whether or not distinguishable particles are in an entangled or untangled quantum state depends on the process that led to the formation of the state.

The wave function for two identical particles must have definite exchange symmetry, and for particles with energy $E_n + E_n'$ this wave function is necessarily an entangled wave function. It can be a symmetric wave function of the form

$$\Psi^{(S)}(x_p, x_q, t) = \frac{1}{\sqrt{2}}[\psi_n(x_p)\psi_{n'}(x_q) + \psi_n(x_q)\psi_{n'}(x_p)]\, e^{-i(E_n + E_{n'})t/\hbar} \tag{10.12}$$

or an antisymmetric wave function of the form

$$\Psi^{(A)}(x_p, x_q, t) = \frac{1}{\sqrt{2}}[\psi_n(x_p)\psi_{n'}(x_q) - \psi_n(x_q)\psi_{n'}(x_p)]\, e^{-i(E_n + E_{n'})t/\hbar}. \tag{10.13}$$

Both these wave functions represent states in which each particle is equally associated with both single-particle states.

We shall now use these wave functions to show that exchange symmetry leads to a surprising tendency for identical particles to either huddle together or to avoid each other. To do so, we shall set $x_p = x_q = x_0$ and compare the values of the wave functions for distinguishable and for identical particles when the particles have the same location x_0. For distinguishable particles with untangled wave functions given by Eqs. (10.9) and (10.10), we obtain

$$\Psi_{1,2}^{(D)}(x_0, x_0, t) = \psi_n(x_0)\psi_{n'}(x_0)\, e^{-i(E_n + E_{n'})t/\hbar}.$$

For identical particles with a symmetrical entangled wave function (10.12) we obtain

$$\Psi^{(S)}(x_0, x_0, t) = \sqrt{2}\psi_n(x_0)\psi_{n'}(x_0)\,e^{-i(E_n+E_n')t/\hbar},$$

and for identical particles with an antisymmetrical entangled wave function (10.13) we obtain

$$\Psi^{(A)}(x_0, x_0, t) = 0.$$

These equations show that, all other things being the equal, two identical particles with a symmetric wave function are twice as likely to be found at the same location as two distinguishable particles with an untangled wave function, and that two identical particles with an antisymmetric wave function are never found at the same location. The wave-mechanical origin of this behaviour is interference, either constructive in a symmetrical wave function or destructive in an antisymmetrical wave function.

The tendency of identical particles for togetherness or for avoidance may be illustrated more fully by considering an example where the particles occupy harmonic oscillator states with $n = 0$ and $n' = 1$. If we use the eigenfunctions listed in Table 6.1 and use the coordinates

$$x = x_p - x_q \quad \text{and} \quad X = \frac{x_p + x_q}{2},$$

we obtain the following expressions for the symmetric and antisymmetric wave functions given by Eqs. (10.12) and (10.13):

$$\Psi^{(S)}(x, X, t) = \frac{2}{a^2\sqrt{\pi}}\,e^{-x^2/4a^2}\,X\,e^{-X^2/a^2}\,e^{-i(E_0+E_1)t/\hbar}$$

and

$$\Psi^{(A)}(x, X, t) = \frac{-1}{a^2\sqrt{\pi}}x\,e^{-x^2/4a^2}\,e^{-X^2/a^2}\,e^{-i(E_0+E_1)t/\hbar}.$$

The square modulus of each of these wave functions gives a probability density for the particles to have a separation x and a centre of mass located at X. By integrating over all possible values of X, we can find the probability for a separation x. Straightforward calculus shows that the probability for a separation with a magnitude between $|x|$ and $|x + dx|$ is

$$P^{(S)}(x)\,dx = \frac{2}{a\sqrt{2\pi}}\,e^{-x^2/2a^2}\,dx \tag{10.14}$$

for the symmetric wave function and

$$P^{(A)}(x)\,dx = \frac{2}{a\sqrt{2\pi}}\frac{x^2}{a^2}\,e^{-x^2/2a^2}\,dx \qquad (10.15)$$

for the antisymmetric wave function. The corresponding probability for two distinguishable particles with an untangled wave function like that given by Eq. (10.9) or by Eq. (10.10) is

$$P^{(D)}(x)\,dx = \frac{1}{a\sqrt{2\pi}}\left(1+\frac{x^2}{a^2}\right)e^{-x^2/2a^2}\,dx. \qquad (10.16)$$

The probability densities $P^{(S)}(x)$, $P^{(A)}(x)$ and $P^{(D)}(x)$ are plotted in Fig. 10.1. The graphs on the left show that identical particles huddle together if they have a symmetric wave function and the graphs on the right show that identical

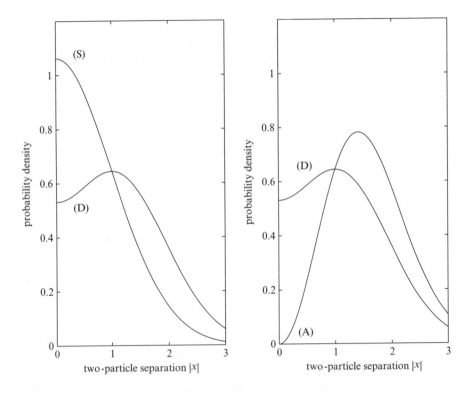

Fig. 10.1 The separation of two identical particles in a one-dimensional harmonic oscillator potential. We have assumed that the particles are in single-particle states with quantum numbers $n=0$ and $n'=1$ and that the harmonic oscillator length parameter a is equal to one. The probability densities $P^{(S)}(x)$ and $P^{(A)}(x)$ for symmetric and antisymmetric wave functions are labelled (S) and (A), and they are compared with the probability density labelled (D) for two distinguishable particles with an untangled wave function given by Eq. (10.9) or by Eq. (10.10).

particles avoid each other if they have an antisymmetric wave function. Similar behaviour could be demonstrated by distinguishable particles, but only when particular physical processes lead to the formation of symmetrical or antisymmetrical entangled states.

We emphasise that these tendencies for togetherness or avoidance are not a consequence of attractive or repulsive forces between the particles; indeed in our illustrative example, the particles are governed by a Hamiltonian, Eq. (10.7), in which there is no direct interaction between the particles. These tendencies are a result of the exchange symmetry that always exists for identical quantum particles. We also note from Fig. 10.1, that the effects of exchange symmetry become less important as the separation of the particles increases. This reflects the fact that if identical particles can keep their distance they become almost distinguishable.

10.3   EXCHANGE SYMMETRY WITH SPIN

We have shown that the indistinguishability of identical particles leads to wave functions with definite exchange symmetry. We have yet to consider the circumstances that determine whether the wave function is symmetric or antisymmetric. These circumstances depend on the spin of the particles, but to understand the role of spin in determining exchange symmetry, we need to extend our representation of a quantum state so that it includes a description of particle spin.

As stated in Section 8.1, a particle is said to have spin s if the magnitude and the z component of its intrinsic angular momentum are given by

$$S = \sqrt{s(s+1)}\hbar \text{ and } S_z = m_s\hbar, \text{ where } m_s = \begin{cases} +s \\ +(s-1) \\ \vdots \\ -(s-1) \\ -s. \end{cases}$$

For example, electrons, protons and neutrons have spin $s = \frac{1}{2}$, the deuteron nucleus ^{2}H has spin $s = 1$ and the helium nucleus ^{4}He has spin $s = 0$.

From Section 8.3 we recall that all the orbital angular momentum properties of a particle p with orbital angular momentum $\sqrt{l(l+1)}\hbar$ can be described using $2l+1$ eigenfunctions $Y_{l,m_l}(\theta_p, \phi_p)$. Similarly, all the intrinsic angular momentum properties of a particle p with spin $\sqrt{s(s+1)}\hbar$ can be described using $2s+1$ eigenvectors $\chi_{s,m_s}(p)$ with $m_s = -s, -s+1, \ldots, +s$. In particular, in analogy with Eq. (8.28), the most general spin state has the form

$$\chi(p) = \sum_{m_s=-s}^{s} c_{m_s} \chi_{s,m_s}(p)$$

where $|c_{m_s}|^2$ is equal to the probability that the particle has a z component of spin equal to $m_s\hbar$. (See footnote 2 below.)

When the spatial and spin properties of a particle are independent of each other, the quantum state may be represented by a product

$$\Phi(p) = \psi(\mathbf{r}_p)\chi(p).$$

For example, the first term $\psi(\mathbf{r}_p)$ describing the spatial properties of the particle could be a hydrogen–like wave function with quantum numbers n, l, m_l and the second term $\chi(p)$ could be a spin state with quantum numbers $s = \frac{1}{2}$ and $m_s = \pm\frac{1}{2}$. When this is the case, we have a single-particle quantum state of the form

$$\Phi_{n, l, m_l, m_s}(p) = \psi_{n, l, m_l}(\mathbf{r}_p)\chi_{s, m_s}(p).$$

We can now write down expressions for quantum states which describe the spatial and spin properties of two identical particles.

When both particles occupy the same single-particle state, say one with spatial and spin quantum numbers n, l, m_l, m_s, we can construct a symmetrical state for two identical particles of the form

$$\Phi^{(S)}(p, q) = \Phi_{n, l, m_l, m_s}(p)\Phi_{n, l, m_l, m_s}(q). \tag{10.17}$$

But an antisymmetric two-particle state for two identical particles cannot be constructed when both particles occupy the same single-particle state. This implies that, when identical particles have antisymmetric exchange symmetry, two or more particles cannot occupy the same single-particle state.

When the particles are associated with two different single-particle states, it is possible to construct both symmetric and antisymmetric two-particle states. For example, we can have a symmetric state of the form

$$\Phi^{(S)}(p, q) = \frac{1}{\sqrt{2}}[\Phi_{n, l, m_l, m_s}(p)\Phi_{n', l', m'_l, m'_s}(q) \\ + \Phi_{n, l, m_l, m_s}(q)\Phi_{n', l', m'_l, m'_s}(p)] \tag{10.18}$$

---

[2] In general, the probability amplitude c_{m_s} depends on time, but to keep the notation simple we shall ignore time dependence. The spin eigenvectors $\chi_{s, m_s}(p)$ are less abstract if they are represented by column matrices with $2s + 1$ components; for example, a particle with spin $s = \frac{1}{2}$ can be described using the matrices

$$\chi_{\frac{1}{2}, +\frac{1}{2}}(p) = \begin{pmatrix} 1 \\ 0 \end{pmatrix}_p \quad \text{and} \quad \chi_{\frac{1}{2}, -\frac{1}{2}}(p) = \begin{pmatrix} 0 \\ 1 \end{pmatrix}_p,$$

where the subscript p is necessary because a matrix representing particle p must be distinguished from a matrix representing another particle. The mathematics of the representation of spin quantum states is covered in more advanced books, but this mathematics will not be needed here.

and an antisymmetric state of the form

$$\Phi^{(A)}(p, q) = \frac{1}{\sqrt{2}} [\Phi_{n, l, m_l, m_s}(p)\Phi_{n', l', m'_l, m'_s}(q)$$
$$- \Phi_{n, l, m_l, m_s}(q)\Phi_{n', l', m'_l, m'_s}(p)]. \tag{10.19}$$

Alternatively, we can construct symmetric and antisymmetric two-particle states by combining two-particle wave functions and two-particle spin states with the appropriate symmetry. We shall illustrate this procedure by considering two spin-half particles.

Spin states for two spin-half particles may be labelled by the quantum numbers S and M_S, indicating that the magnitude and z component of the combined spin angular momentum of the state are $\sqrt{S(S+1)}\hbar$ and $M_S\hbar$, respectively.[3] We have already encountered the rule for the addition of the orbital and spin angular momenta of a single particle in Section 9.6. The rule for the addition of two spins is very similar. In particular, two spins with quantum number $s = \frac{1}{2}$ can be combined to give a combined spin with quantum number

$$S = \tfrac{1}{2} + \tfrac{1}{2} = 1 \quad \text{or} \quad S = \tfrac{1}{2} - \tfrac{1}{2} = 0.$$

This rule is merely a quick way of saying that two spin-half states, $\chi_{\frac{1}{2}, m_s}(p)$ and $\chi_{\frac{1}{2}, m_s}(q)$, may be combined to give two-particle spin states with quantum numbers $S = 1$ and $S = 0$. Explicitly, if we use a simplified notation in which $\chi_{\frac{1}{2}, m_s}$ is denoted by χ_+ when $m_s = +\frac{1}{2}$ and χ_- when $m_s = -\frac{1}{2}$, we have three symmetric spin states,

$$\left.\begin{aligned}
\chi^{(S)}_{1, +1}(p, q) &= \chi_+(p)\chi_+(q) \\
\chi^{(S)}_{1, 0}(p, q) &= \frac{1}{\sqrt{2}}[\chi_+(p)\chi_-(q) + \chi_+(q)\chi_-(p)] \\
\chi^{(S)}_{1, -1}(p, q) &= \chi_-(p)\chi_-(q),
\end{aligned}\right\} \tag{10.20}$$

which correspond to $S = 1$ and $M_S = +1, 0$ and -1, and one antisymmetric spin state

$$\chi^{(A)}_{0, 0}(p, q) = \frac{1}{\sqrt{2}}[\chi_+(p)\chi_-(q) - \chi_+(q)\chi_-(p)] \tag{10.21}$$

which corresponds to $S = 0$ and $M_S = 0$.

---

[3] We follow the convention of using the capital letters S and M_S to denote the spin angular momentum of two or more electrons, even though the symbol S was also used to denote the magnitude of a spin angular momentum in Chapter 8.

These two-particle spin states may be combined with two-particle wave functions to produce a two-particle quantum state with definite exchange symmetry. For example, we can construct antisymmetric quantum states for two electrons, in two ways. We can combine a symmetric spin state with quantum numbers $S = 1$ and $M_S = 1, 0, -1$ and an antisymmetric wave function $\psi^{(A)}(\mathbf{r}_p, \mathbf{r}_q)$ to give

$$\Phi^{(A)}(p, q) = \psi^{(A)}(\mathbf{r}_p, \mathbf{r}_q)\chi^{(S)}_{S, M_S}(p, q), \tag{10.22}$$

or we can combine an antisymmetric spin state with quantum numbers $S = 0$ and $M_S = 0$ and a symmetric wave function $\psi^{(S)}(\mathbf{r}_p, \mathbf{r}_q)$ to give

$$\Phi^{(A)}(p, q) = \psi^{(S)}(\mathbf{r}_p, \mathbf{r}_q)\chi^{(A)}_{S, M_s}(p, q). \tag{10.23}$$

We have chosen this example because, as we shall discover in the next section, electron quantum states are always antisymmetric. This means that the symmetry of the wave function of two electrons is fixed by the value of the combined spin of the electrons in accordance with Eqs. (10.22) and (10.23).

Finally, in order to keep this and the preceding section as simple as possible, we have only discussed quantum states for two identical particles. When several identical particles are considered, similar conditions for acceptable quantum states emerge. They are that the quantum state of a system of identical particles must be either symmetric or antisymmetric when any two of the particles are exchanged.

10.4   BOSONS AND FERMIONS

According to the arguments presented in the last section, a quantum state describing the spatial and spin properties of identical particles must have definite exchange symmetry and there are two possible options for this symmetry: symmetric or antisymmetric. But, in the real world, identical particles do not have the freedom to choose between these options. In the real world there are two types of quantum particles called *bosons* and *fermions* with the following characteristics:

- Bosons are particles with integer spin and a system of identical bosons must have quantum states which are symmetric when two particles are exchanged.

- Fermions are particles with half-integer spin and systems of identical fermions must have quantum states which are antisymmetric when two particles are exchanged.

This surprising connection between the exchange symmetry and the spin of identical particles is called the *spin-statistics theorem*. Its far-reaching consequences for the properties of matter were identified in the early days of quantum mechanics and it was subsequently shown to be a theoretical consequence of relativistic quantum field theory. The spin-statistics theorem means that in applied quantum mechanics we encounter two ways of being indistinguishable, the boson way and the fermion way.

Because electrons have spin half, they are indistinguishable in the fermion way. They never have symmetric states like Eq. (10.18) but only antisymmetric states like Eq. (10.19). These antisymmetric states for electrons are very relevant to our understanding of the world around us because the properties of atoms and solids are largely determined by the quantum mechanics of electrons. The three most important consequences of the antisymmetry of electron quantum states are the following:

First, a single-particle state can be occupied by at most one electron, because, if it were occupied by more than one electron, the multi-electron quantum state would be symmetric and not antisymmetric when the electrons were exchanged. This characteristic of electrons is called the *Pauli exclusion principle*.[4] The Pauli principle plays a governing role in determining the physical and chemical properties of atoms and leads, as we shall show in the next chapter, to an understanding of *the Periodic Table of the Elements*. It also has a crucial role in determining the electrical and thermal properties of electrons in solids.

Second, when the spin part of the quantum state of two electrons is symmetric, the spatial part, i.e. the wave function, is necessarily antisymmetric as in Eq. (10.22). When this is the case, the electrons, like the two identical particles in a harmonic oscillator potential described by the graph on the right of Fig. 10.1, have a strong tendency to avoid each other. This tendency for avoidance is ultimately responsible for the rigidity of ordinary matter; a solid resists compression because identical electrons avoiding each other prevent atoms from getting closer.

Third, when the spin state of two electrons is antisymmetric, the wave function is necessarily symmetric as in Eq. (10.23). In this case, the electrons, like the two identical particles described by the graph on the left of Fig. 10.1, have a tendency to huddle together. When this happens between adjacent atoms, a covalent bond between the atoms may be created and a molecule formed.

Protons and neutrons, like electrons, are also spin-half fermions, and the fermion-like indistinguishability of protons and of neutrons plays a governing role in the shell model of the nucleus. In this model protons and neutrons occupy single-particle states, but a single-particle state can be occupied by at

---

[4] We note with admiration that W Pauli discovered the exclusion principle empirically in 1924 and that he also proved the spin-statistics theorem some 17 years later using relativistic quantum field theory.

most one proton and by one neutron. Similarly, theoretical models of protons, neutrons and other hadrons are governed by the idea that quarks of a specific flavour and colour also act like systems of identical fermions with antisymmetric quantum states.

The boson way of being indistinguishable also leads to important physical phenomena. Because bosons are described by symmetric quantum states, many bosons may occupy the same single-particle state and when this happens quantum-mechanical behaviour on a macroscopic scale may arise.

The most important example of boson togetherness is the coherent light of a laser. This coherence arises because photons, bosons with spin one, have a high probability to have the same energy and momentum, in much the same way as two particles with a symmetric wave function have a high probability of being at the same location.

Boson togetherness is also responsible for the superfluidity of liquid helium at temperatures below 2.2 K. Liquid helium consists of a system of weakly interacting helium atoms which behave like bosons because they consist of a ^{4}He nucleus with spin zero and two electrons with a combined spin of zero. At low temperatures, a considerable fraction of the atoms in liquid helium 'condense' into the same lowest-energy state. They form a *Bose–Einstein condensate* in which the atoms have wave functions which are coherent with each other and move collectively without friction. Recently almost pure Bose–Einstein condensates have been produced by cooling atoms in magnetic traps; indeed the 2001 Nobel Prize in Physics was awarded to Eric Cornell, Wolfgang Ketterle and Carl Wieman for their work in producing the first pure Bose–Einstein condensate in 1995.

Surprisingly, boson-like togetherness also occurs in situations where fermion-like behaviour is expected. It occurs in the superconductivity of metals at low temperatures because pairs of electrons act like indistinguishable bosons. It also probably occurs when liquid helium-3 becomes a superfluid at very low temperatures. Helium-3 atoms, unlike the normal helium atoms, are fermions because the ^{3}He nucleus has spin half, but pairs of helium-3 atoms can act like a system of indistinguishable bosons and give rise to collective motion with no friction in liquid helium-3.

PROBLEMS 10

1. In Section 10.1 we explained why the wave function of two identical particles has a definite exchange symmetry. In this problem we show that this exchange symmetry remains unchanged as the wave function evolves.

   Given that the time evolution of the wave function for two particles is governed by the Schrödinger equation,

$$i\hbar \frac{\partial \Psi(\mathbf{r}_p, \mathbf{r}_q, t)}{\partial t} = \hat{H}(\mathbf{r}_p, \mathbf{r}_q)\Psi(\mathbf{r}_p, \mathbf{r}_q, t),$$

where $\hat{H}(\mathbf{r}_p, \mathbf{r}_q)$ is the Hamiltonian operator for particles p and q, show that the relation between the wave functions at time t and at time $t + dt$, is

$$\Psi(\mathbf{r}_p, \mathbf{r}_q, t + dt) = \left[1 - \frac{i}{\hbar} \hat{H}(\mathbf{r}_p, \mathbf{r}_q) dt\right] \Psi(\mathbf{r}_p, \mathbf{r}_q, t).$$

What general symmetry condition must be satisfied by the Hamiltonian operator $\hat{H}(\mathbf{r}_p, \mathbf{r}_q)$ in order that the exchange symmetry of the wave function is the same at times t and $t + dt$? Explain why this condition is always satisfied when the particles are identical.
(Hint: Eq. (10.7) is an example of a two-particle Hamiltonian operator.)

2. The wave function $\Psi(x_p, x_q, t)$ of two particles with mass m and total energy E in a one-dimensional harmonic oscillator potential has the form

$$\Psi(x_p, x_q, t) = \psi(x_p, x_q) e^{-iEt/\hbar},$$

where $\psi(x_p, x_q)$ satisfies the two-particle eigenvalue equation

$$\left[-\frac{\hbar^2}{2m}\frac{\partial^2}{\partial x_p^2} - \frac{\hbar^2}{2m}\frac{\partial^2}{\partial x_q^2} + \frac{1}{2}m\omega^2 x_p^2 + \frac{1}{2}m\omega^2 x_q^2\right]\psi = E\psi.$$

Show by substitution that the function

$$\psi(x_p, x_q) = \psi_n(x_p)\psi_{n'}(x_q),$$

where ψ_n and $\psi_{n'}$ are one-particle harmonic oscillator eigenfunctions with energies E_n and $E_{n'}$, satisfies these equations if $E = E_n + E_{n'}$.

3. Consider the following wave function for two distinguishable particles

$$\Psi^{(D)}(x_p, x_q, t) = \psi_n(x_p)\psi_{n'}(x_q) e^{-i(E_n+E_{n'})t/\hbar},$$

and the following wave functions for two identical particles

$$\Psi^{(S)}(x_p, x_q, t) = \frac{1}{\sqrt{2}}[\psi_n(x_p)\psi_{n'}(x_q) + \psi_n(x_q)\psi_{n'}(x_p)] e^{-i(E_n+E_{n'})t/\hbar}$$

and

$$\Psi^{(A)}(x_p, x_q, t) = \frac{1}{\sqrt{2}}[\psi_n(x_p)\psi_{n'}(x_q) - \psi_n(x_q)\psi_{n'}(x_p)] e^{-i(E_n+E_{n'})t/\hbar}.$$

(a) Show that all of these wave functions are normalized if the single-particle eigenfunctions ψ_n and $\psi_{n'}$ are normalized and orthogonal.

(b) Suppose that there is a weak, repulsive interaction between the particles given by the potential $V(|x_p - x_q|)$ which causes the energy of particles with wave function $\Psi(x_p, x_q, t)$ to shift by

$$\Delta E = \int_{-\infty}^{+\infty} dx_p \int_{-\infty}^{+\infty} dx_q \; \Psi^*(x_p, x_q, t) V(|x_p - x_q|) \Psi(x_p, x_q, t).$$

How does the energy shift for distinguishable particles compare with the energy shifts for identical particles?

4. Consider two non-interacting particles p and q each with mass m in a cubical box of size a. Assume that the energy of the particles is

$$E = \frac{3\hbar^2 \pi^2}{2ma^2} + \frac{6\hbar^2 \pi^2}{2ma^2}.$$

Using the eigenfunctions

$$\psi_{n_x, n_y, n_z}(x_p, y_p, z_p) \quad \text{and} \quad \psi_{n_x, n_y, n_z}(x_q, y_q, z_q)$$

given by Eq. (4.43) in Section 4.4, write down two-particle wave functions which could describe the system when the particles are:

(a) distinguishable, spinless bosons;

(b) identical, spinless bosons;

(c) identical spin-half fermions in a symmetric spin state;

(d) identical spin-half fermions in the antisymmetric spin state.

5. Suppose that five non-interacting particles are placed in a three-dimensional harmonic oscillator potential described in Section 6.5, for which the single-particle energy is

$$E_n = \left(n_x + n_y + n_z + \tfrac{3}{2}\right)\hbar\omega.$$

What is the lowest energy of the five-particle state when the particles are:

(a) distinguishable, spinless bosons;

(b) identical, spinless bosons;

(c) identical fermions each with spin $s = \tfrac{1}{2}$;

(d) identical fermions each with spin $s = \tfrac{3}{2}$?

6. The hydrogen molecule contains two identical, spin-half nuclei which can vibrate and rotate. We discussed the vibrational energies in Section 6.4 and the rotational energies in problem 4 at the end of Chapter 8. We shall label the nuclei by p and q, and describe their spin properties using the spin states $\chi_{S,M_S}(p, q)$ with $S = 1, M_S = 1, 0, -1$ and $S = 0, M_S = 0$. When the molecule is in a rotational state with orbital angular momentum quantum numbers l and m_l, the spatial separation of the nuclei $\mathbf{r} = \mathbf{r}_p - \mathbf{r}_q$ is governed by a wave function of the form

$$\psi_{l,m_l}(\mathbf{r}) = R(r)\, Y_{l,m_l}(\theta, \phi).$$

Bearing in mind the symmetry properties of the spin states given in Eqs. (10.20) and (10.21) and the symmetry properties of the spherical harmonics given in Table 8.1, explain why the rotational states of the hydrogen molecule with odd l have nuclei with a combined spin given by $S = 1$ and those with even l have nuclei with a combined spin given by $S = 0$.

(Because the spins of the nuclei in a hydrogen molecule are seldom affected by a collision, a gas of hydrogen acts as if it consists of two types of molecules, *ortho-hydrogen* with rotational states with odd l and *para-hydrogen* with rotational states with even l. It takes days for a gas of ortho-hydrogen to reach thermal equilibrium with a gas of para-hydrogen.)

11

Atoms

As Richard Feynman once said, why do chemists count funny? Instead of saying $1, 2, 3, 4, \ldots$, they say hydrogen, helium, lithium, beryllium, ... The reason is remarkable. An atom with Z electrons is identical to all other atoms with Z electrons. They have the same size, the same ionization energy and the same tendency to react or not to react. Atomic sameness may be labelled by one number, the atomic number Z, and there are only 100 or so different types. Moreover, this sameness is resilient. When an atom is excited by an encounter with a photon, an electron or another atom, it will always return to its original pristine condition.

The resilient sameness of atoms with atomic number Z cannot be understood in terms of classical physics. It is a property of the quantum states of Z electrons. The essential features of these quantum states may be understood by combining, in an approximate way, the concepts we used to describe the hydrogen atom with the implications that arise from the fact that electrons are indistinguishable fermions. Even though we have to be content with approximate representations for the quantum states of an atom with atomic number Z, these states in principle provide a complete description which is unique to all atoms with atomic number Z.

11.1 ATOMIC QUANTUM STATES

An atom with atomic number Z consists of Z electrons held together in the potential energy field due to the Coulomb attraction of a nucleus and the Coulomb repulsion between each pair of electrons. In the helium atom, for example, two electrons move in the potential energy field,

$$V(\mathbf{r}_p, \mathbf{r}_q) = -\frac{2e^2}{4\pi\epsilon_0 r_p} - \frac{2e^2}{4\pi\epsilon_0 r_q} + \frac{e^2}{4\pi\epsilon_0 |\mathbf{r}_p - \mathbf{r}_q|}, \qquad (11.1)$$

and the energy levels and eigenfunctions can be found by solving the eigenvalue equation

$$\left[-\frac{\hbar^2}{2m_e}(\nabla_p^2 + \nabla_q^2) + V(\mathbf{r}_p, \mathbf{r}_q)\right]\psi(\mathbf{r}_p, \mathbf{r}_q) = E\psi(\mathbf{r}_p, \mathbf{r}_q). \qquad (11.2)$$

Accurate numerical solutions of this equation may be obtained, but for atoms with many electrons, approximate methods based on *the central field approximation* must be used. These methods, which were first developed by E. Fermi, D. R. Hartree and L. H. Thomas, lead to the following qualitative description of atomic quantum states.

The central field approximation

In the central field approximation, each electron in an atom moves independently in a central potential due to the Coulomb attraction of the nucleus and the average effect of the other electrons in the atom. If the electron under consideration has coordinate r_p, it will see, when r_p is large, a nucleus with charge Ze screened by inner electrons with total charge $-(Z-1)e$ and we expect a potential energy of the form

$$V_B(r_p) = -\frac{e^2}{4\pi\epsilon_0 r_p}. \qquad (11.3)$$

But when r_p is small, the electron sees a bare, unscreened nucleus and we expect a potential energy of the form

$$V_C(r_p) = -\frac{Ze^2}{4\pi\epsilon_0 r_p}. \qquad (11.4)$$

These considerations imply that the rough form of the central potential for each independent electron in an atom is similar to the curve (A) shown Fig. 11.1. At large distances this potential approaches the Coulomb potential due to a point charge e, curve (B), and at small distances it approaches the Coulomb potential due to a point charge Ze, curve (C). A very simple model for this central potential is given by

$$V_A(r_p) = -\frac{z(r_p)e^2}{4\pi\epsilon_0 r_p}, \quad \text{where} \quad z(r_p) = (Z-1)e^{-r/a} + 1. \qquad (11.5)$$

This equation with a equal to half a Bohr radius and $Z = 6$ was used to calculate curve (A) in Fig. 11.1.

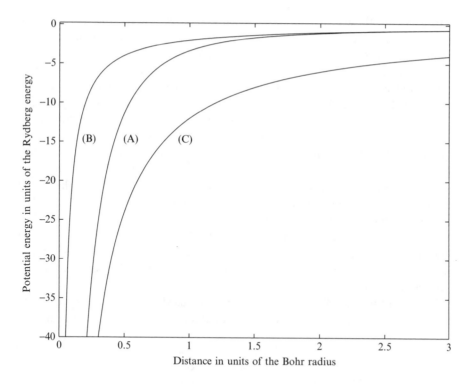

Fig. 11.1   A simple model for the central potential for an electron in a carbon atom with atomic number $Z = 6$. Curve (A) gives the screened potential energy $V_A(r)$ given by Eq. (11.5) in units of the Rydberg energy E_R as a function of r in units of the Bohr radius, a_0, with a screening radius $a = a_0/2$. Curves (B) and (C) give the potentials due to point charges e and $6e$, $V_B(r)$ and $V_C(r)$ given by Eqs. (11.3) and (11.4).

Given a central potential $V(r)$ one can find the energy levels and eigenfunctions for each electron in the atom by solving the single-particle energy eigenvalue equation

$$\left[-\frac{r^2}{2m_e}\nabla^2 + V(r)\right]\psi(\mathbf{r}) = E\psi(\mathbf{r}). \tag{11.6}$$

This equation is identical to Eq. (9.5) which was the starting point for our discussion in Chapter 9 of the quantum mechanics of a particle in a central potential and of the hydrogen atom. This means that the eigenfunctions for these single-particle states may be labelled by the same quantum numbers used to label the eigenfunctions of the hydrogen atom: the principal quantum number n, the orbital angular quantum numbers l and m_l, and the electron spin quantum number m_s. But in contrast with Eq. (9.22) for the hydrogen atom, the energy levels depend on two of these quantum numbers, n and l.

The energy of these levels may be labelled E_{nl} but we will use the spectroscopic notation 1s, 2s, 2p, etc. We shall also follow the atomic physics convention of referring to these single-particle states as *orbitals*.

The effect of screening on the energies of the orbitals in a carbon atom is illustrated in Fig. 11.2. The 1s, 2s and 2p energy levels for the unscreened potential $V_C(r)$ are shown on the left and those for the screened potential $V_A(r)$ are shown on the right. We note that for the screened potential, the energy of an orbital increases when the principal quantum number n increases, and that, for a given value of n, the energy increases when the orbital angular momentum quantum number l increases. This increase with l can be understood by recalling from Section 9.1 that the effective potential in the radial Schrödinger equation includes the centrifugal potential, $l(l+1)\hbar^2/2m_e r^2$. Because the centrifugal potential becomes more repulsive as l increases, an electron with a higher value for l has a lower probability of penetrating close to the nucleus and a higher probability of being located at larger distances where the nucleus is well screened by inner electrons. The net effect of higher l is a weaker attraction by the nucleus and a higher energy.

The eigenfunctions for these orbitals may be used to construct approximate multi-electron quantum states for the atom as a whole. As discussed in Chapter

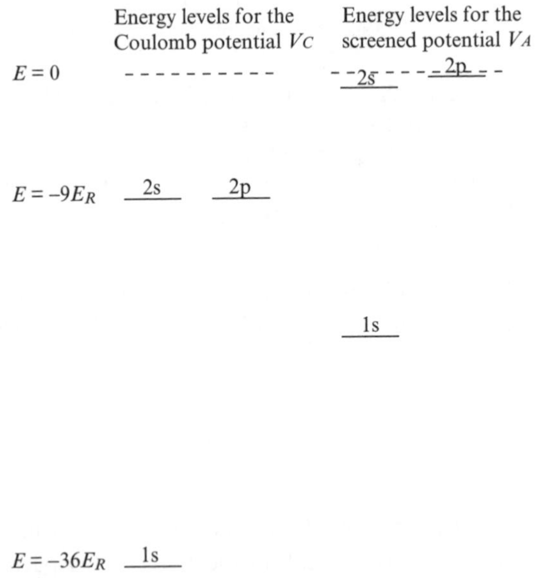

Energy levels for the   Energy levels for the
Coulomb potential V_C screened potential V_A

$E = 0$ $-\,-\,-\,-\,-\,-\,-\,-\,-\,-$ $-\,-\underline{2s}\,-\,-\,-\,-\underline{2p}\,-\,-$

$E = -9E_R$ $\underline{\quad 2s \quad}$ $\underline{\quad 2p \quad}$

$\underline{\quad 1s \quad}$

$E = -36E_R$ $\underline{\quad 1s \quad}$

Fig. 11.2 The energies of the 1s, 2s and 2p orbitals in the carbon atom. The effect of electron screening is to raise the energy levels. This can be seen by comparing the energy levels on the left, given by an unscreened Coulomb potential $V_C(r)$ due a point charge with $Z = 6$, with those on the right given by the screened potential $V_A(r)$ with $Z = 6$ and a equal to half a Bohr radius; the energies for the screened potential were calculated numerically by Daniel Guise.

10, quantum states which describe a system of indistinguishable electrons must be antisymmetric whenever two electrons are exchanged. This can only be achieved if electrons are assigned to orbitals in accordance with the Pauli exclusion principle; i.e. not more than one electron may occupy an orbital with the same quantum numbers n, l, m_l, m_s. This means that at most two electrons can be assigned to 1s orbitals, one with quantum numbers $n = 1$, $l = 0$, $m_l = 0$, and $m_s = +\frac{1}{2}$, and one with quantum numbers $n = 1$, $l = 0$, $m_l = 0$, and $m_s = -\frac{1}{2}$. Similarly, not more than two electrons can be assigned to 2s orbitals, but up to six electrons can be assigned to 2p orbitals because there are six of these orbitals with quantum numbers $n = 2$, $l = 1$, $m_l = +1, 0, -1$, and $m_s = \pm\frac{1}{2}$. When the 1s, 2s and 2p orbitals are fully occupied, additional electrons may only be assigned to orbitals with principal quantum numbers n greater than 2. These orbitals have higher energy and also a limited capacity.

We can illustrate how to use this construction kit for atomic states by considering a carbon atom containing six electrons which may occupy energy levels similar to those shown on the right-hand-side of Fig. 11.2. The ground state is obtained by assigning six electrons to orbitals with the lowest possible energy; a maximum of two electrons can have the energy E_{1s}, a maximum of two can have the energy E_{2s} and the minimum energy for each of the two remaining electrons is E_{2p}. These assignments give an *electron configuration* denoted by $(1s)^2(2s)^2(2p)^2$ with energy

$$E = 2E_{1s} + 2E_{2s} + 2E_{2p}.$$

The first excited state of the carbon atom is obtained by assigning only one electron to a 2s orbital and three electrons to 2p orbitals. This gives rise to the electron configuration $(1s)^2(2s)(2p)^3$ with energy

$$E = 2E_{1s} + E_{2s} + 3E_{2p}.$$

If the energy levels shown in Fig. 11.2 for the screened potential $V_A(r)$ are used as a rough guide, the energy of the ground state is $-41.1E_R$ and the energy of the first excited state is $-40.6E_R$. Clearly, states of higher excitation may be obtained by assigning more electrons to 2p orbitals or by assigning electrons to 3s, 3p, 3d, ... orbitals.

In the preceding paragraph we have followed custom and given the wrong impression that particular electrons are in particular orbitals. This is not the case. Because all the electrons in the atom are indistinguishable, each electron is equally associated with each of the occupied orbitals. In fact, like the two-electron state given by Eq. (10.19), the multi-electron quantum state is antisymmetric when any two of the electrons are exchanged.

We have also wrongly given the impression that the central potential which represents the effect of the attraction of the nucleus and of the average effects of electron–electron repulsion, is easy to find. In fact, the central potential and the eigenfunctions for independent electrons in an atom are like the chicken and the egg; it is unclear which comes first. This problem may be resolved by a sequence of calculations in which a central potential gives rise to eigenfunctions and electron probabilities which in turn can be used to find an improved approximation to the central potential. Ultimately a self-consistent central potential emerges which accurately describes independent electrons in the atom.

Corrections to the central field approximation

There are two important corrections to the central field approximation. The first is due to *a residual electron–electron repulsion* which arises because the Coulomb repulsion between the electrons in the atom cannot be described accurately by a central potential. The second is due to *a spin–orbit interaction* which is similar to that considered for the hydrogen atom in Section 9.6. But before we can understand these corrections, we need to develop a more precise description of atomic quantum states which takes into account how the angular momenta of atomic electrons may be combined. *Readers who are content with an approximate description based solely on electron configurations should proceed directly to Section 11.2.*

Because the angular momenta of atomic electrons involve unfamiliar notation, we shall begin by recalling some basic facts and stating some conventions about angular momentum.

- The magnitude and z component of an angular momentum can have the values

$$\sqrt{j(j+1)}\hbar \quad \text{and} \quad m_j\hbar,$$

where j is a quantum number which can equal $0, \frac{1}{2}, 1, \frac{3}{2}, \ldots$ and m_j is a quantum number which can equal $j, j-1, \ldots, -j$.

- Two angular momenta with quantum numbers j_1 and j_2 may be combined to give an angular momentum with quantum number j which can take on the values

$$j = j_1 + j_2, \ j_1 + j_2 - 1, \ldots, |j_1 - j_2|.$$

- Lower case letters are used in atomic physics for the angular momentum of a single electron: l and m_l for orbital angular momentum, s and m_s for spin angular momentum, and j and m_j for a combined orbital and spin angular momentum. But capital letters are used for the angular momentum of two or more electrons: L and M_L for orbital angular momentum, S and M_S for spin angular momentum, and J and M_J for a combined orbital and spin angular momentum.

- Spectroscopic notation uses the letters s, p, d, f...to denote the orbital angular momentum of a single electron with $l = 0, 1, 2, 3$.... But capital letters S, P, D, F...are used to denote the orbital angular momentum of two or more electrons with $L = 0, 1, 2, 3$....

With these preliminaries out of the way, we can describe how the angular momenta of atomic electrons may be combined. Two ways of combining angular momenta are used in atomic physics. They are called $L - S$ coupling (also called Russell-Saunders coupling) and $j - j$ coupling.

In $L - S$ coupling, the orbital angular momenta of the electrons are coupled to give a combined orbital angular momentum described by a quantum number L, and the spin angular momenta of the electrons are coupled to give a combined spin angular momentum described by a quantum number S. These combined orbital and spin angular momenta are then coupled to give a total angular momentum described by a quantum number J. In $j - j$ coupling, the orbital and spin angular momenta of each electron are coupled to give a combined angular momentum described by a quantum number j. The combined angular momenta of each of the electrons are then coupled to give a total angular momentum described by a quantum number J. The $L - S$ coupling scheme is most useful when the residual electron–electron repulsion is larger than the spin–orbit interaction, and $j - j$ coupling is most useful in the converse situation. In practice, $j - j$ coupling is used for atoms with high atomic numbers and $L - S$ coupling is used for atoms with low atomic numbers.

We shall illustrate these general ideas by considering how $L - S$ coupling can be used to describe the carbon atom. We shall focus on the low-lying quantum states with the electron configuration $(1s)^2 (2s)^2 (2p)^2$. As stressed in Section 10.4, these quantum states must be antisymmetric when any two electrons are exchanged.

The two electrons in the 1s orbitals, which necessarily have a symmetric wave function, are in an antisymmetric quantum state if they have an antisymmetric spin state. Such a spin state, according to Eq. (10.21), has zero spin. Thus, in addition to having a zero orbital angular momentum, the two 1s electrons have a combined spin angular momentum equal to zero. This is also the case for the two 2s electrons.

The two electrons in the 2p orbitals, each with orbital angular momentum quantum number $l = 1$, can have a combined orbital angular momentum with a quantum number L which can take on the values 2, 1 or 0, and a combined spin angular momentum with a quantum number S which can take on the values 1 or 0. As shown in Eqs. (10.20) and (10.21), the spin state with $S = 1$ is symmetric and the spin state with $S = 0$ is antisymmetric. It can be also be shown that the exchange symmetry of the orbital angular momentum wave functions is symmetric when $L = 2$, antisymmetric when $L = 1$ and symmetric when $L = 0$. Hence, acceptable quantum states with antisymmetric exchange symmetry only exist for the two 2p electrons if the spin and orbital angular momentum quantum numbers S and L occur in one of the combinations

$$\text{(i) } S = 1, \ L = 1; \quad \text{(ii) } S = 0, \ L = 2; \quad \text{(iii) } S = 0, \ L = 0.$$

These states are usually denoted by the spectroscopic notation ^{3}P, ^{1}D and ^{1}S where P denotes $L = 1$, D denotes $L = 2$ and S denotes $L = 0$, and the superscript denotes the number of possible values for the z component of the spin, $2S + 1$. (States with $2S + 1 = 1$ and $2S + 1 = 3$ are called singlet and triplet states respectively.) Because there are also $2L + 1$ possible values of the z component of the orbital angular momentum, there are $3 \times 3\,^3$P states, $1 \times 5\,^1$D states and $1 \times 1\,^1$S states, giving a total of 15 possible states with the electron configuration $(1s)^2\,(2s)^2\,(2p)^2$. If the electrons in an atom really moved in a central potential, all these 15 states would have the same energy

$$E = 2E_{1s} + 2E_{2s} + 2E_{2p}.$$

But this degeneracy does not occur because not all of the effects of electron–electron repulsion can be included in the central potential. There is a residual electron–electron repulsion which causes the quantum states ^{3}P, ^{1}D and ^{1}S to have different energy levels, as shown in Fig. 11.3. These energy differences arise because the effect of the residual electron–electron repulsion is different in wave functions with different exchange symmetry and different orbital angular momentum.

   In addition to the residual electron–electron repulsion, a spin–orbit interaction, similar to that given by Eq. (9.35) for the hydrogen atom in Section 9.6, also affects the energy levels of an atom. It gives rise to a *fine structure* in which the energy depends, not only on the configuration and on the values L and S, but also on the value of the total angular momentum quantum number J, which can take on values between

$$J = L + S \quad \text{and} \quad J = |L - S|.$$

For example, if we consider the ^{3}P states of the carbon atom we can couple the orbital and spin angular momenta with $L = 1$ and $S = 1$ to give states with a total angular momentum with J equal to 2, 1 or 0. These states are usually denoted by ^{3}P$_2$, ^{3}P$_1$ and ^{3}P$_0$, where the subscript denotes the value of the quantum number J, and they have slightly different energies, as illustrated on the right-hand side of Fig.11.3. The energy levels of the ^{1}D and ^{1}S states are not split by the spin–orbit interaction because only one value of J is possible for these states, $J = 2$ and $J = 0$ respectively.

   To sum up, we have discussed how the energy levels of the carbon atom may be described with increasing precision. To a first approximation, the energy of the level is determined by an electron configuration which describes independent electrons in a central potential which represents the Coulomb attraction of

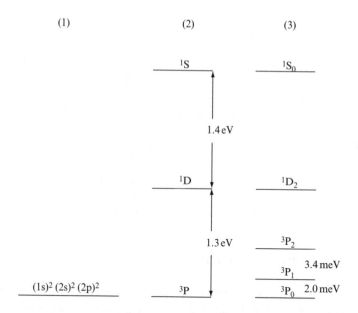

Fig. 11.3 Three approximations for the lowest energy levels of the carbon atom. (1) A single level given by the energy of the configuration $(1s)^2 (2s)^2 (2p)^2$. (2) Three levels split by the residual electron–electron repulsion, labelled by the quantum numbers L and S. (3) Five levels depending on the spin–orbit interaction, labelled by the quantum numbers L, S and J; the energy differences of the $^3P_0$, $^3P_1$ and $^3P_2$ levels, which are of the order of milli-electron volts, have not been drawn to scale.

the nucleus and the average effect of the Coulomb repulsion between electrons in the atom. To a second approximation, the energy depends upon the residual electron–electron repulsion and the level is labelled by the orbital and spin quantum numbers L and S. To a third approximation, the energy depends upon the spin–orbit interaction and the level is labelled by L, S and by the total angular momentum quantum number J.

Finally, we note that most of our knowledge about atomic energy levels is derived from the spectral lines that arise when radiative transitions occur between the levels. The most probable transitions for emission or absorption in the visible and the ultra-violet region of the spectrum are electric dipole transitions, which were discussed in the simpler context of the hydrogen atom in Section 9.4. For atoms more complex than the hydrogen atom, the selection rules for electric dipole transitions are more complicated than the $\Delta l = \pm 1$ rule given in Eq. (9.28). The change in the total angular momentum quantum number, $\Delta J = J_f - J_i$, obeys the rule

$$\Delta J = 0, \pm 1, \quad \text{but} \quad J_i = 0 \rightarrow J_f = 0 \text{ is strictly forbidden.} \tag{11.7}$$

When $L - S$ coupling provides an accurate description of the quantum states, the following rules are also obeyed

$$\Delta L = 0, \pm 1 \quad \text{and} \quad \Delta S = 0. \tag{11.8}$$

There is an additional selection rule, as mentioned in Section 9.4: the *parity* of an atomic state must change in an electric dipole transition. The parity is determined by the behaviour of the quantum state when all the coordinates undergo a reflection through the origin. If the state is unchanged it has even parity and if the state is changed by a factor of -1 it has odd parity. As stated at the end of Section 9.1, the parity of a one-electron state is even when the orbital angular momentum quantum number l is even and it is odd when l is odd. More generally, if an atomic state has a single configuration $(n_1 l_1)(n_2 l_2)\ldots(n_Z l_Z)$, its parity is even if the sum of the orbital angular quantum numbers $l_1 + l_2 + \ldots + l_Z$ is even and it is odd if this sum is odd.

11.2   THE PERIODIC TABLE

Atoms with different atomic numbers Z have different chemical properties, but these properties vary periodically with Z. For example, lithium, sodium, potassium and rubidium with $Z = 3$, 11, 19 and 37, respectively, are called alkali metals because they have similar metallic properties; fluorine, chlorine and bromine with $Z = 9$, 17, and 35, respectively, are called halogens because they have similar reactive properties; and helium, neon, argon and krypton with $Z = 2$, 10, 18 and 36, respectively, are called noble gases because they show little tendency to react. In fact, every chemical element belongs to a group of elements with similar properties and each of these groups contains elements with a well-defined sequence of atomic numbers.

Quantum mechanics provides a physical explanation for why the properties of the chemical elements vary periodically. Only two quantum mechanical ideas are involved. The first is that each electron in an atom occupies a single-particle state, or orbital, with definite energy. The second is that electrons occupy these orbitals in accordance with the Pauli exclusion principle.

The orbitals occupied by atomic electrons have energy levels which depend upon the value of the atomic number Z of the atom; with minor variations for atoms with atomic numbers greater than 20, the sequence of the energy levels is the same for all atoms. This sequence is shown in Fig. 11.4. We note that, as in Fig.11.2, the energy of an orbital increases when the principal quantum number n increases, and that, for a given value of n, the energy increases when the orbital angular momentum quantum number l increases. An important consequence of the increase of energy with l is that the energies of orbitals with different values of n may be close together. In particular, the 3d and 4s energies are close, the 4d and 5s energies are close and the 4f, 5d and 6s energies are close.

Given the sequence of energy levels shown in Fig. 11.4 and the constraints imposed by the Pauli exclusion principle, we can write down electron configurations for any chemical element. The lowest-energy electron configurations for

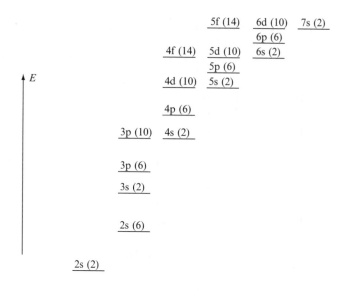

Fig. 11.4 A schematic representation of the sequence of energy levels for electron orbitals in an atom. The number in the brackets is the degeneracy of the level which equals the maximum number of electrons that can be assigned to the level. The order of the levels depends, to a limited extent, on the atomic number Z of the atom. For example, the 4s level is lower than the 3d level for calcium with $Z = 20$, but the 4s level is higher than the 3d level for scandium with $Z = 21$.

the first 30 chemical elements, together with their ionization energies, are given in Table 11.1. This table shows how a many-electron atom, in its ground state, has a specific set of occupied orbitals. Because an orbital with higher energy has a greater spatial extent, the atom has several *shells* of negative charge. For example, an argon atom at $Z = 18$ has a K shell formed by occupied 1s orbitals, an L shell formed by occupied 2s and 2p orbitals and an M shell formed by occupied 3s and 3p orbitals.

As we move down Table 11.1, we periodically encounter atoms with common properties for their least bound electrons.

Most importantly, we encounter, at $Z = 2$, 10, and 18, helium, neon and argon atoms with a closed-shell structure. These are very tightly bound systems with small sizes and high ionization energies. They are also difficult to excite; the minimum excitation energy is about 20 eV for helium, 16 eV for neon and 10 eV for argon. These atoms are almost chemically inert.

At $Z = 3$, 11 and 19 we encounter lithium, sodium and potassium atoms with a structure consisting of tightly bound closed shells of electrons plus one loosely

TABLE 11.1.   The lowest-energy electron configurations and ionization energies E_I of the first 30 elements

Element		Z	Electron configuration	E_I (eV)
Hydrogen	H	1	$(1s)$	13.6
Helium	He	2	$(1s)^2$	24.6
Lithium	Li	3	$(1s)^2(2s)$	5.4
Beryllium	Be	4	$(1s)^2(2s)^2$	9.3
Boron	B	5	$(1s)^2(2s)^2(2p)$	8.3
Carbon	C	6	$(1s)^2(2s)^2(2p)^2$	11.3
Nitrogen	N	7	$(1s)^2(2s)^2(2p)^3$	14.5
Oxygen	O	8	$(1s)^2(2s)^2(2p)^4$	13.6
Fluorine	F	9	$(1s)^2(2s)^2(2p)^5$	17.4
Neon	N	10	$(1s)^2(2s)^2(2p)^6$	21.6
Sodium	Na	11	$(1s)^2(2s)^2(2p)^6(3s)$	5.1
Magnesium	Mg	12	$(1s)^2(2s)^2(2p)^6(3s)^2$	7.6
Aluminium	Al	13	$(1s)^2(2s)^2(2p)^6(3s)^2(3p)$	6.0
Silicon	Si	14	$(1s)^2(2s)^2(2p)^6(3s)^2(3p)^2$	8.1
Phosphorus	P	15	$(1s)^2(2s)^2(2p)^6(3s)^2(3p)^3$	10.5
Sulfur	S	16	$(1s)^2(2s)^2(2p)^6(3s)^2(3p)^4$	10.4
Chlorine	Cl	17	$(1s)^2(2s)^2(2p)^6(3s)^2(3p)^5$	13.0
Argon	A	18	$(1s)^2(2s)^2(2p)^6(3s)^2(3p)^6$	15.8
Potassium	K	19	$(1s)^2(2s)^2(2p)^6(3s)^2(3p)^6(4s)$	4.3
Calcium	Ca	20	$(1s)^2(2s)^2(2p)^6(3s)^2(3p)^6(4s)^2$	6.1
Scandium	Sc	21	$(1s)^2(2s)^2(2p)^6(3s)^2(3p)^6(3d)(4s)^2$	6.5
Titanium	Ti	22	$(1s)^2(2s)^2(2p)^6(3s)^2(3p)^6(3d)^2(4s)^2$	6.8
Vanadium	V	23	$(1s)^2(2s)^2(2p)^6(3s)^2(3p)^6(3d)^3(4s)^2$	6.7
Chromium	Cr	24	$(1s)^2(2s)^2(2p)^6(3s)^2(3p)^6(3d)^5(4s)$	6.8
Manganese	Mn	25	$(1s)^2(2s)^2(2p)^6(3s)^2(3p)^6(3d)^5(4s)^2$	7.4
Iron	Fe	26	$(1s)^2(2s)^2(2p)^6(3s)^2(3p)^6(3d)^6(4s)^2$	7.9
Cobalt	Co	27	$(1s)^2(2s)^2(2p)^6(3s)^2(3p)^6(3d)^7(4s)^2$	7.9
Nickel	Ni	28	$(1s)^2(2s)^2(2p)^6(3s)^2(3p)^6(3d)^8(4s)^2$	7.6
Copper	Cu	29	$(1s)^2(2s)^2(2p)^6(3s)^2(3p)^6(3d)^{10}(4s)$	7.7
Zinc	Zn	30	$(1s)^2(2s)^2(2p)^6(3s)^2(3p)^6(3d)^{10}(4s)^2$	6.7

bound electron. These atoms are called alkali metal atoms and they react readily with other atoms, normally by transferring their loosely bound, or *valence* electron.

At $Z = 9$ and 17 we encounter fluorine and chlorine atoms with a structure which needs one more electron to form a closed shell. These atoms have high ionization energies but they are reactive because the acquisition of an extra electron is energetically favourable; the energy released in binding an extra electron, *the electron affinity*, is 3.5 eV for fluorine and 3.6 eV for chlorine.

Finally, we note that the periodicity of the periodic table becomes somewhat erratic for atoms with higher atomic numbers. This behaviour sets in at $Z = 19$ because the 3d and 4s energy levels are very close together and for some values of Z the 3d level is higher than the 4s level and for other values of Z it is not, as demonstrated by the electron configurations of elements between potassium and zinc given in Table 11.1. Similar complications arise when the 4d and 5s orbitals are filled, when the 4f, 5d and 6s orbitals are filled, and when the 5f, 6d and 7s orbitals are filled, but we will leave these complications to chemists to sort out.

11.3   WHAT IF?

It should be clear from the last section that the periodic variation of the properties of atoms with atomic number is largely a consequence of the Pauli exclusion principle, a principle that arises because electrons are indistinguishable quantum particles which conceal their identity by having antisymmetric quantum states. Because the Pauli principle is a symmetry principle which has no analogue in classical physics, we shall end this chapter by reflecting on what atoms would be like without it. We shall do so by using an approximate and simple model which is based on the intuition we should have acquired by working through 240 or so pages of quantum mechanics. We shall introduce the model by considering two atoms whose ground states are unaffected by the Pauli principle, the hydrogen and helium atoms.

The hydrogen atom consists of one electron and a nucleus of charge e. If the atom is in a quantum state with spatial extent R, the average potential energy is of the order of

$$V = -\frac{e^2}{4\pi\epsilon_0 R}$$

and the average kinetic energy is at least of the order of

$$T = \frac{\hbar^2}{2m_e R^2}.$$

This expression for the kinetic energy can be thought of as the minimum kinetic energy of an electron localized in a region of size R. It can be derived by noting that the average momentum of the electron is comparable with the uncertainty in its momentum, Δp, and that, in accordance with the uncertainty principle, the minimum value of this uncertainty is of the order of

$$\Delta p = \frac{\hbar}{R}.$$

The sum of the kinetic and potential energies gives a total energy of the order of

$$E = \frac{\hbar^2}{2m_e R^2} - \frac{e^2}{4\pi\epsilon_0 R}. \tag{11.9}$$

For future reference, we will rewrite this energy as

$$E = \frac{A_1}{R^2} - \frac{B_1}{R}, \tag{11.10}$$

where A_1 and B_1 are the constants

$$A_1 = \frac{\hbar^2}{2m_e} \quad \text{and} \quad B_1 = \frac{e^2}{4\pi\epsilon_0}, \tag{11.11}$$

which determine the kinetic energy and the potential energy for a state of the hydrogen atom with spatial extent R. We note that for a state with large R, the energy is dominated by the potential energy $-B_1/R$, that for a state with small R, the energy is dominated by the kinetic energy of localization A_1/R^2, and that a balance between the attractive effect of the potential energy and the repulsive effect of the kinetic energy gives rise to a minimum energy when

$$\frac{dE}{dR} = -2\frac{A_1}{R^3} + \frac{B_1}{R^2} = 0,$$

i.e. when $R = 2A_1/B_1$. We conclude that this model predicts a ground state with an energy and size given by

$$E_1 = -\frac{B_1^2}{4A_1} \quad \text{and} \quad R_1 = \frac{2A_1}{B_1}. \tag{11.12}$$

When we substitute for A_1 and B_1 and use the expressions for the Bohr radius a_0 and Rydberg energy E_R, Eqs. (9.19) and (9.20), we find that Eq. (11.12) gives the correct energy and radius for the ground state of the hydrogen atom,

$$E_1 = -E_R \quad \text{and} \quad R_1 = a_0. \tag{11.13}$$

We shall now show that the model, with minor adjustments, can also describe the energy and size of the ground state of the helium atom. In this atom there are two electrons and a nucleus of charge $2e$. If both electrons are in the same single-particle state, the energy of a two-electron quantum state of size R is roughly given by

$$E = 2\frac{\hbar^2}{2m_e R^2} - 4\frac{e^2}{4\pi\epsilon_0 R} + \frac{e^2}{4\pi\epsilon_0 R_{ee}}, \tag{11.14}$$

where the last term represents the potential energy due to Coulomb repulsion between two electrons with separation R_{ee}. Because the typical electron–electron separation is of the order of R, we shall approximate the potential energy of repulsion by setting

$$R_{ee} = fR \qquad (11.15)$$

where f is an electron–electron avoidance parameter; we expect f to be of the order of one, but greater than one if the electrons are good at avoiding each other. With this approximation, the energy of the two-electron atom becomes

$$E = 2\frac{\hbar^2}{2m_e R^2} - \left(4 - \frac{1}{f}\right)\frac{e^2}{4\pi\epsilon_0 R}. \qquad (11.16)$$

In analogy with our description of the hydrogen atom, we rewrite this expression as

$$E = \frac{A_2}{R^2} - \frac{B_2}{R}, \quad \text{with} \quad A_2 = 2A_1 \quad \text{and} \quad B_2 = \left(4 - \frac{1}{f}\right)B_1, \qquad (11.17)$$

where A_1 and B_1 are the constants given by Eq. (11.11) that determine the energy of the hydrogen atom.

Thus, in the helium atom, a balance is struck between the attractive and repulsive effects of the potential energy $-B_2/R$ and the kinetic energy of localization A_2/R^2 to give a quantum state with minimum energy $E_2 = -B_2^2/4A_2$ and size $R_2 = 2A_2/B_2$. If we substitute for A_2 and B_2 we obtain

$$E_2 = -\frac{\left(4 - \frac{1}{f}\right)^2}{2}E_R \quad \text{and} \quad R_2 = \frac{2}{\left(4 - \frac{1}{f}\right)}a_0. \qquad (11.18)$$

This formula for E_2 gives the correct binding energy of the ground state of helium, $5.8E_R$, if we take the electron–electron avoidance parameter to be 1.67, which in turn gives a spatial extent of $R_2 = 0.6a_0$, that is in agreement with the measured radius of $0.6a_0$.

Having established an acceptable model for the ground states of hydrogen and helium, we will use the model to describe the ground state of a hypothetical atom containing Z electrons which are not constrained by the Pauli exclusion principle. When all of the Z electrons are in the same single-particle state, the total energy is given by

$$E = Z\frac{\hbar^2}{2m_e R^2} - Z^2\frac{e^2}{4\pi\epsilon_0 R} + \frac{Z(Z-1)}{2}\frac{e^2}{4\pi\epsilon_0 R_{ee}}. \qquad (11.19)$$

If we assume that the separation of each of the $Z(Z - 1)/2$ pairs of electrons is given by $R_{ee} = fR$, we obtain

$$E = Z\frac{\hbar^2}{2m_e R^2} - \left(Z^2 - \frac{Z(Z-1)}{2f}\right)\frac{e^2}{4\pi\epsilon_0 R}, \qquad (11.20)$$

which can be rewritten as

$$E = \frac{A_Z}{R^2} - \frac{B_Z}{R}, \quad \text{with} \quad A_Z = ZA_1 \quad \text{and} \quad B_Z = \left(Z^2 - \frac{Z(Z-1)}{2f}\right)B_1. \quad (11.21)$$

By finding the minimum of this energy, we obtain the following expressions for the energy and size of the ground state of our hypothetical atom:

$$E_Z = -\frac{\left(Z^2 - \frac{Z(Z-1)}{2f}\right)^2}{Z}E_R \quad \text{and} \quad R_Z = \frac{Z}{\left(Z^2 - \frac{Z(Z-1)}{2f}\right)}a_0. \qquad (11.22)$$

To find the ionization energy of the atom, i.e. the energy needed to remove one electron, we need to know the energy of $Z - 1$ electrons in a negative ion with a nucleus of charge Ze. By working through problem 8 at the end of this chapter, you can easily show that this energy is given by

$$E_{Z-} = -\frac{\left(Z(Z - 1) - \frac{(Z-1)(Z-2)}{2f}\right)^2}{(Z - 1)}E_R. \qquad (11.23)$$

The ionization energy may then be found from

$$E_I = E_{Z-} - E_Z. \qquad (11.24)$$

But before we can do so, we need to specify the value of the electron–electron avoidance parameter f. Even though we expect f to decrease with Z, we shall for simplicity use the same value, $f = 1.67$, which worked for the helium atom. Straightforward algebra then gives

$$E_I = (0.07Z^2 + 0.57Z + 0.36)E_R \quad \text{and} \quad R_Z = \left(\frac{1}{0.7Z + 0.3}\right)a_0. \qquad (11.25)$$

We can now explicitly illustrate the role of the Pauli exclusion principle in atoms. We shall do this by comparing the predictions of Eq. (11.25) for a hypothetical atom with atomic number Z with the measured ionization energy and radius of the real atom with atomic number Z. The results of this comparison for the first 11 elements are shown in Fig 11.5. The solid circles give energies and radii for real atoms in which the Pauli principle has a governing role, and the open circles give the energies and radii for hypothetical atoms in which the Pauli principle plays no role.

We see that without the Pauli principle, the ionization energies steadily increase and the radii steadily decrease with atomic number Z. In particular, the periodicity of chemical properties of real atoms is replaced by a chemistry in which atoms steadily become less reactive than helium. A world without the Pauli exclusion principle would be very different. One thing is for certain: it would be a world with no chemists.

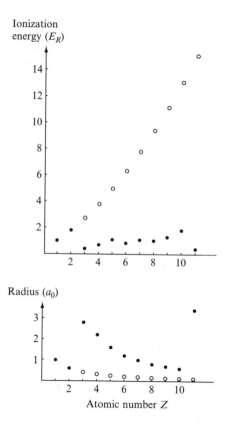

Fig. 11.5 The effect of the Pauli exclusion principle on the ionization energies and radii of the first 11 elements. The solid circles correspond to atoms in which the Pauli principle constrains the behaviour of the electrons and the hollow circles correspond to atoms in which the constraints of the Pauli principle are not imposed. The Pauli principle has no effect on the ground states of the first elements, hydrogen and helium. The Rydberg energy and the Bohr radius have been used as units for the ionization energies and radii.

PROBLEMS 11

1. Assume that $Z - 1$ electrons around a nucleus with charge Ze have a charge density

$$\rho(r) = -\frac{(Z-1)e}{4\pi a^3}\frac{e^{-r/a}}{r/a}.$$

   (a) Show that the mean charge radius of this distribution of electrons is equal to $2a$.

   (b) Find the charge inside a radius r.

   (c) Show that an electron moving in the electrostatic field of the nucleus and of this distribution of electrons has a potential energy given by Eq. (11.5), i.e.

$$V(r) = -\frac{z(r_p)e^2}{4\pi\epsilon_0 r}, \quad \text{where} \quad z(r) = (Z-1)e^{-r/a} + 1.$$

   (Hint: For part (c) use Gauss' theorem to find the magnitude of the electric field $E(r)$ and find the electrostatic potential $\phi(r)$ using $E = -d\phi/dr$.)

2. Given that the energy needed to remove the two electrons from the ground state of the helium atom is 79 eV, show that the energy needed to remove just one electron from the ground state of the helium atom is 24.6 eV.

   (In this problem and in the next problem you may use without derivation the expression you obtained in problem 14 at the end of Chapter 9 for the energy of an electron moving in the Coulomb potential of a point charge.)

3. Consider what chemists call the valence electron in a sodium atom. The other 10 electrons partially screen the charge $11e$ on the nucleus so that, to a first approximation, the valence electron sees an effective point charge Z^*e which is less than $11e$. The binding energy of the valence electron in sodium is 5.12 eV for the 3s state, 2.10 eV for the 3p state and 1.52 eV for the 3d state. What are the magnitudes of the effective point charge, Z^*, of the screened nucleus as seen by an electron in a 3s, in a 3p and in a 3d state?

4. Explain why up to 10 electrons may be assigned to 3d orbitals in an atom and why 14 can be assigned to 4f orbitals.

5. Consider the following electron configurations of the helium atom:

$(1s)^2$, $(1s)(2s)$, $(1s)(2p)$ and $(1s)(3d)$.

Write down the possible values for the total orbital angular momentum quantum number L and the possible values for the total spin quantum number S. For each value for L and S write down the possible values for the total angular momentum quantum number J.

6. Consider the following electron configurations of the carbon atom:

$$(1s)^2(2s)^2(2p)(3s) \quad \text{and} \quad (1s)^2(2s)^2(2p)(3p).$$

Write down the possible values for the total orbital angular momentum quantum number L and the possible values for the total spin quantum number S. For each value of L and S write down the possible values for the total angular momentum quantum number J.

7. Explain why there is a ^3S state of the carbon atom with the electron configuration $(1s)^2(2s)^2(2p)(3p)$ but not one with the configuration $(1s)^2(2s)^2(2p)^2$.

8. Consider an ion with $Z - 1$ electrons and a nucleus with charge Ze. Using the model described in Section 11.3, write down an expression for the energy of a quantum state of the ion with spatial extent R. Show that the minimum of this energy is

$$E_{Z^-} = -\frac{\left(Z(Z-1) - \frac{(Z-1)(Z-2)}{2f}\right)^2}{(Z-1)} E_R.$$

9. According to the model used in Section 11.3, the energy of a two-electron quantum state with spatial extent R of an ion with a nucleus with charge Ze is

$$E = 2\frac{\hbar^2}{2m_e R^2} - 2Z\frac{e^2}{4\pi\epsilon_0 R} + \frac{e^2}{4\pi\epsilon_0 R_{ee}}, \quad \text{with} \quad R_{ee} = fR.$$

(a) Consider the two-electron ions with the experimental ionization energies, E_I^{expt} in units of the Rydberg energy, given below. Using the electron–electron avoidance parameter that worked for the helium atom, $f = 1.67$, show that the model predicts the theoretical ionization energies given by E_I^{theory} in the following Table.

Ion	Li^+	Be^{2+}	B^{3+}	C^{4+}	N^{5+}	O^{6+}	F^{7+}
E_I^{expt}	5.57	11.32	19.08	28.84	40.61	54.39	70.12
E_I^{theory}	5.6	11.4	19.2	29.0	40.8	54.6	70.4

(b) A hydrogen atom is capable of binding a second electron to form a H^- ion. Show that the model gives the experimental binding energy of this ion, 0.75 eV, if the electron–electron avoidance parameter f is equal to 1.8.

Hints to selected problems

CHAPTER 1

1. Rewrite the equations for the conservation of momentum and of energy in the form

$$|\mathbf{p}_i - \mathbf{p}_f| = |\mathbf{P}_f - \mathbf{P}_i| \quad \text{and} \quad \epsilon_i - \epsilon_f = E_f - E_i,$$

and make use of the relations between the energy and momentum of a particle. Finally, make use of the relation between the energy of a photon and its wavelength.

2. If the minimum energy needed to eject an electron is W, the photon must have a frequency which is at least W/h.

3. Because kT is an energy and hc is an energy-length, $(kT)^4/(hc)^3$ is an energy per unit volume and $c(kT)^4/(hc)^3$ is an energy per second per unit area. Hence, the energy radiated from unit area of a black-body with temperature T is expected to be

$$I = \sigma T^4 = f\,\frac{c(kT)^4}{(hc)^3}$$

where f is a dimensionless constant.

4. In the theory of relativity $m_e c^2$ is the rest energy of an electron and in electromagnetism $e^2/4\pi\epsilon_0 R$ is a potential energy if R is a length. It follows that

$$R = \frac{e^2}{4\pi\epsilon_0 m_e c^2}$$

is a fundamental length in a relativistic theory of the electron. Show that this length is $\alpha^2 a_0$ where a_0 is the Bohr radius and $\alpha = e^2/4\pi\epsilon_0\hbar c$. The constant α is called the fine structure constant and it is approximately equal to $1/137$.

5. The force between the electron and the proton, $e^2/4\pi\epsilon_0 r^2$, causes a centripetal acceleration equal to $m_e v^2/r$. The orbital angular momentum of the electron is $L = mvr$.

6. Make use of the fact that the magnitude of the momentum of a particle is at least as big as the uncertainty in its momentum and use the uncertainty principle Eq. (1.15).

7. Use the uncertainty principle to show that the uncertainty in the momentum of the quark, and hence the minimum value of its average momentum, is small compared with mc where m is the mass of the quark.

8. Evaluate the de Broglie wavelength of electrons with kinetic energy $200\,\mathrm{eV}$ and consider the condition for strong diffraction by a slit.

9. Show that the de Broglie wavelength of a $54\,\mathrm{eV}$ electron is $\lambda = 0.166\,\mathrm{nm}$. The condition for constructive interference of waves scattered by atoms on the surface is $D\sin\phi = n\lambda$. Show that this condition is satisfied when $D = 0.215\,\mathrm{nm}$, $\phi = 50\,$degrees and $n = 1$.

10. Show that the wave due to a conduction electron in copper is strongly diffracted by the lattice of atoms. To do this show that the de Broglie wavelength of a $7\,\mathrm{eV}$ electron is $0.46\,\mathrm{nm}$ and that this is comparable with the distance between atoms in copper.

11. Show that the de Broglie wavelength of a neutron with thermal energy $\frac{3}{2}kT$ is comparable with the distance between atoms in a solid if $T = 300\,\mathrm{K}$.

12. Estimate the thermal energy of an oxygen molecule at $T = 273\,\mathrm{K}$ and show that the de Broglie wavelength is much smaller than the typical distance between molecules in air.

CHAPTER 2

1. The phase and group velocities are given by

$$v_{phase} = \frac{\omega}{k} \quad \text{and} \quad v_{group} = \frac{d\omega}{dk}.$$

Verify that $v_{group} = \frac{3}{2} v_{phase}$.

2. Use

$$\int \cos k'(x - ct)\, dk' = \frac{\sin k'(x - ct)}{(x - ct)}$$

and

$$\sin A - \sin B = 2\cos\left(\frac{A+B}{2}\right)\sin\left(\frac{A-B}{2}\right).$$

3. Use relations like

$$\frac{\partial \cos(kx - \omega t)}{\partial t} = \omega \sin(kx - \omega t)$$

and

$$\frac{\partial^2 \cos(kx - \omega t)}{\partial x^2} = -k^2 \cos(kx - \omega t).$$

4. The expression

$$\Psi(x,\ t) = 2iA \sin kx\ \mathrm{e}^{-i\omega t}$$

is obtained by using $\mathrm{e}^{+i\theta} - \mathrm{e}^{-i\theta} = 2i\sin\theta$. It is a complex standing wave with wave number k and angular frequency ω.

5. Verify that the wave function is a solution of the differential equation

$$-i\hbar\frac{\partial\Psi}{\partial t} = -\frac{\hbar^2}{2m}\frac{\partial^2\Psi}{\partial x^2}.$$

6. (a) Use

$$v_{group} = \frac{d\omega}{dk}.$$

(b) Show that $\Psi = A\,\mathrm{e}^{-i(\omega t - kx)}$ is a solution if $\epsilon^2 - p^2c^2 = m^2c^4$.

CHAPTER 3

1. Make use of

$$\mathrm{e}^{-a}\sum_{n=0}^{\infty}\frac{a^n}{n!} = \mathrm{e}^{-a}\,\mathrm{e}^{+a} = 1.$$

2. (a) Evaluate the integral

$$\int_{-\infty}^{+\infty} x\rho(x)\,\mathrm{d}x$$

by sketching the integrand $x\rho(x)$.

(b) Show that

$$\frac{d\rho}{dx} = -\frac{x}{\sigma^2}\rho$$

and find $\langle x^2 \rangle$ by using

$$\int_{-\infty}^{+\infty} x^2 \rho(x)\,dx = -\sigma^2 \int_{-\infty}^{+\infty} x\frac{d\rho}{dx}\,dx$$

and the identity given in the question.

(c) Use $\sigma = \sqrt{\langle x^2 \rangle - \langle x \rangle^2}$.

3. (a) Evaluate the integral

$$\int_0^\infty p(t)\,dt.$$

(b) Evaluate the integral

$$\int_0^\infty tp(t)\,dt.$$

(c) The probability for living for at least time T is equal to the probability for decay between $t = T$ and $t = \infty$ and this probability is given by

$$\int_T^\infty p(t)\,dt.$$

4. (a) Show that the position probability density is

$$|\Psi(x)|^2 = N^2 e^{-x^2/2\sigma^2}, \quad \text{where} \quad 2\sigma^2 = a^2.$$

Use the properties of the Gaussian distribution to evaluate the sandwich integrals for $\langle x \rangle$ and $\langle x^2 \rangle$.

(b) Use the properties of the Gaussian distribution to evaluate the sandwich integrals for $\langle p \rangle$ and $\langle p^2 \rangle$.

(c) Use

$$\Delta x = \sqrt{\langle x^2 \rangle - \langle x \rangle^2} \quad \text{and} \quad \Delta p = \sqrt{\langle p^2 \rangle - \langle p \rangle^2}.$$

5. (a) Show that

$$\int_0^a |\Psi(x)|^2 \, dx = \frac{N^2 a^5}{30}.$$

(b) Show that

$$\langle x \rangle = \frac{a}{2} \quad \text{and} \quad \langle x^2 \rangle = \frac{2a^2}{7}.$$

Show that

$$\langle p \rangle = 0 \quad \text{and} \quad \langle p^2 \rangle = \frac{10\hbar^2}{a^2}.$$

7. (a) Show that $|\Psi(x)|^2$ has a maximum at $x = 2/\alpha$.

(b) Show that

$$\langle x \rangle = \frac{3}{\alpha} \quad \text{and} \quad \langle x^2 \rangle = \frac{12}{\alpha^2}.$$

(c) Show that

$$\langle p \rangle = 0 \quad \text{and} \quad \langle p^2 \rangle = \frac{\hbar^2 \alpha^2}{4}.$$

(d) Show that

$$\Delta x \, \Delta p = \sqrt{\frac{3}{4}} \, \hbar.$$

CHAPTER 4

3. (a) Assume the well width a is equal to the Bohr radius a_0 and note that the Rydberg energy is given by

$$E_R = \frac{\hbar^2}{2m_e a_0^2} = 13.6 \, \text{eV}.$$

(b) Assume the well width a is equal to $1 \, \text{fm} = 1 \times 10^{-15} \, \text{m}$.

4. Rewrite using

$$\sin k_n x = \frac{e^{+ik_n x} - e^{-ik_n x}}{2i}.$$

5. Use the two-dimensional equivalents of Eqs. (4.43) and (4.44).

6. Consider the illustrative example provided by Eq. (4.57) and modify.

8. The wave function may be expressed as

$$\Psi(x,\ 0) = \sum_{n=1}^{\infty} c_n \psi_n(x)$$

where $\psi_n(x)$ is the eigenfunction of the particle with energy E_n in an infinite square well of width a. As in problem 7

$$c_n = \frac{2}{a} \int_0^a \sin k_n x\ \Psi(x,\ 0)\ \mathrm{d}x.$$

Find $|c_1|^2$.

9. By showing that $e^{-iE_n T/\hbar} = 1$ verify that $\Psi(t + T) = \Psi(t)$.

10. The wavelength λ of the emitted radiation, given by

$$\frac{hc}{\lambda} = E_2 - E_1,$$

is uncertain if E_2 has uncertainty $\Delta E_2 = \hbar/\tau$.

CHAPTER 5

1. (a) The eigenvalue equation is given by Eq. (5.5) when $|x| < a$ and by Eq. (5.7) when $|x| > a$.

   (b) and (c) Modify the mathematics leading to Eq. (5.11) and the graphical solution shown in Fig. 5.2.

4. (a) Substitute the expressions for ψ and E into the eigenvalue equation

$$-\frac{\hbar^2}{2m}\frac{\mathrm{d}^2\psi}{\mathrm{d}x^2} + V(x)\psi = E\psi$$

   and find the form of $V(x)$ which ensures that the equation is satisfied.

   (b) The eigenfunction of the first excited state has a node between $x = 0$ and $x = \infty$.

6. The wave function for $x < 0$ is given by Eq. (5.29), for $0 < x < a$ by Eq. (5.31) and for $x > a$ by Eq. (5.34). Impose the conditions that ψ and $d\psi/dx$ are continuous at $x = 0$ and $x = a$.

7. Use Eqs. (5.44), (5.50) and (5.51).

8. The half-life is inversely proportional to the decay rate.

CHAPTER 6

3. Show that if the energy of a classical particle with amplitude A equals $\frac{1}{2}\hbar\omega$ then $A = a = \sqrt{\hbar/m\omega}$.

   Show that the probability for finding the quantum particle in the region $|x| > A$ is

   $$\frac{2}{a\sqrt{\pi}} \int_a^\infty e^{-x^2/a^2} \, dx.$$

4. The potential is identical to that of the harmonic oscillator for $0 < x < \infty$, but presents an infinite barrier which prevents the particle from entering the region $-\infty < x < 0$. The energies are $\frac{3}{2}\hbar\omega$, $\frac{7}{2}\hbar\omega$, etc because the eigenfunctions $\psi(x)$ have the following properties:

   (i) $\psi(x) = 0$ for $-\infty < x < 0$;

   (ii) $\psi(x)$ is identical to a harmonic oscillator eigenfunction for $0 < x < \infty$;

   (iii) $\psi(x)$ is continuous at $x = 0$ but this is only satisfied if $\psi(x)$ in the region $0 < x < \infty$ is identical to a harmonic oscillator eigenfunction with $n = 1, 3, 5, \ldots$

5. (a) According to Eq. (6.15)

   $$\Psi(x, \, t) = \frac{1}{\sqrt{2}}[\psi_0(x)\,e^{-iE_0 t/\hbar} + \psi_1(x)\,e^{-iE_1 t/\hbar}].$$

   (b) Follow the steps that led to Eq. (4.53).

   (c) Use $E_1 - E_0 = \hbar\omega$.

6. (a) and (b) Use Eq. (4.54).

   (c) Use $e^{+i\theta} + e^{-i\theta} = 2\cos\theta$.

7. Use Eq. (6.20) to show that $k = 1548\,\mathrm{N\,m^{-1}}$.

8. (a) and (b) Modify the arguments set out in Section 6.5.

   (c) Consider linear superpositions of the form $\psi_{1,0} \pm i\psi_{0,1}$.

CHAPTER 7

1. Show that

$$[\hat{x}, \hat{p}] = i\hbar, \quad [\hat{x}, \hat{H}] = \frac{\hbar^2}{m} \frac{\partial}{\partial x} \quad \text{and} \quad [\hat{p}, \hat{H}] = -i\hbar \frac{dV}{dx}.$$

2. (d) Show that the functions $A\,e^{\pm ikx}$ are simultaneous eigenfunctions of $\hat{T}$ and $\hat{p}$.

5. (a) The complex conjugate of the expectation value is

$$\langle A \rangle^* = \int_{-\infty}^{+\infty} \Psi\,(\hat{A}\Psi)^* \, dx = \int_{-\infty}^{+\infty} (\hat{A}\Psi)^* \, \Psi \, dx.$$

Because $\hat{A}$ is Hermitian

$$\int_{-\infty}^{+\infty} (\hat{A}\Psi)^* \, \Psi \, dx = \int_{-\infty}^{+\infty} \Psi^* \, \hat{A}\, \Psi \, dx = \langle A \rangle.$$

Hence

$$\langle A \rangle^* = \langle A \rangle.$$

(b) If we rewrite

$$\langle A^2 \rangle = \int_{-\infty}^{+\infty} \Psi^* \, \hat{A}\hat{A}\, \Psi \, dx,$$

using $\Psi_2 = \hat{A}\Psi$ we obtain

$$\langle A^2 \rangle = \int_{-\infty}^{+\infty} \Psi^* \, \hat{A}\, \Psi_2 \, dx.$$

Because $\hat{A}$ is Hermitian we have

$$\langle A^2 \rangle = \int_{-\infty}^{+\infty} (\hat{A}\Psi)^* \, \Psi_2 \, dx = \int_{-\infty}^{+\infty} (\hat{A}\Psi)^* \, (\hat{A}\Psi) \, dx.$$

(c) Because $\hat{A}$ is Hermitian

$$\int_{-\infty}^{+\infty} \Psi^* \, \hat{A}\hat{B}\, \Psi \, dx = \int_{-\infty}^{+\infty} (\hat{A}\Psi)^* \, \hat{B}\, \Psi \, dx.$$

Because $\hat{B}$ is Hermitian

$$\int_{-\infty}^{+\infty} (\hat{A}\Psi)^* \, \hat{B} \, \Psi \, dx = \int_{-\infty}^{+\infty} (\hat{B}\hat{A}\Psi)^* \, \Psi \, dx. = \left(\int_{-\infty}^{+\infty} \Psi^* \, \hat{B}\hat{A} \, \Psi \, dx \right)^*.$$

7. (a) The Hamiltonian commutes with $\hat{p}_x = -i\hbar\partial/\partial x$ if the potential energy field satisfies the condition

$$\frac{\partial V}{\partial x} = 0 \quad \text{for all} \quad x, \, y \quad \text{and} \quad z.$$

This means that the potential energy field is unchanged under the translation $x \rightarrow x + a$.

(b) See hint for part (a).

8. (a) Use

$$\hat{\mathbf{r}} = \mathbf{r} \quad \text{and} \quad \hat{\mathbf{p}} = -i\hbar\nabla.$$

(b) Using the definition of a commutator we have

$$\int \psi_E^* \, [\hat{\mathbf{r}} \cdot \hat{\mathbf{p}}, \, \hat{H}] \, \psi_E \, d^3\mathbf{r} = \int \psi_E^* \, (\hat{\mathbf{r}} \cdot \hat{\mathbf{p}}\hat{H} - \hat{H}\hat{\mathbf{r}} \cdot \hat{\mathbf{p}}) \, \psi_E \, d^3\mathbf{r}.$$

Assuming that $\hat{H}$ is Hermitian we have

$$\int \psi_E^* \, [\hat{\mathbf{r}} \cdot \hat{\mathbf{p}}, \, \hat{H}] \, \psi_E \, d^3\mathbf{r} = \int \psi_E^* \, \hat{\mathbf{r}} \cdot \hat{\mathbf{p}} \, \hat{H} \, \psi_E \, d^3\mathbf{r} - \int (\hat{H}\psi_E)^* \, \hat{\mathbf{r}} \cdot \hat{\mathbf{p}} \, \psi_E \, d^3\mathbf{r}.$$

Because ψ_E is an eigenfunction of $\hat{H}$ with a real eigenvalue E we can use

$$\hat{H}\psi_E = E\psi_E \quad \text{and} \quad (\hat{H}\psi_E)^* = E\psi_E^*$$

to obtain

$$\int \psi_E^* \, [\hat{\mathbf{r}} \cdot \hat{\mathbf{p}}, \, \hat{H}] \, \psi_E \, d^3\mathbf{r} = 0.$$

CHAPTER 8

2. (a) The classical angular momentum, $mvr = 9.1 \times 10^{-33}$ J s, is approximately equal to $\sqrt{l(l+1)}\hbar$ if $l = 86$.

3. (a) The energy splitting arises from the interaction of the electron magnetic moment with the magnetic field. According to Eq. (8.16), the splitting is $2 \times \mu_B B = 5.8 \times 10^{-5}$ eV.

(b) Additional splitting equal to $2 \times 2.79 \mu_N B = 8.8 \times 10^{-8}$ eV arises from the inter-action of the proton magnetic moment with the magnetic field.

4. (a) The classical rotational energy is $L^2/2I$ where L is the angular momentum and I is the moment of inertia about the centre of mass. The moment of inertia is $I = ma^2$ and, in quantum mechanics, the eigenvalues of L^2 are $l(l+1)\hbar^2$ with $l = 0, 1, 2, \ldots$

(b) For every value of l, L_z can have $2l + 1$ values given by $L_z = m_l \hbar$ with $m_l = -l, \ldots +l$. Hence, there are $2l + 1$ independent eigenfunctions with rota-tional energy $E_l = l(l+1)\hbar^2/ma^2$.

(c) For a hydrogen molecule $E_1 - E_0 = 1.5 \times 10^{-2}$ eV. Note that the rotational states of the hydrogen molecule are excited at room temperature because $kT \approx 1/40$ eV.

5. (a) Show that

$$\hat{L}_z Z_{m_l} = m_l \hbar Z_{m_l}.$$

(b) A point with coordinates (r, θ, ϕ) also has coordinates $(r, \theta, \phi + 2\pi)$.

6. Use

$$\sin 2\phi = \frac{e^{+i2\phi} - e^{-i2\phi}}{2i} \quad \text{and} \quad \cos \phi = \frac{e^{+i\phi} + e^{-i\phi}}{2},$$

and show that

$$\psi(r, \theta, \phi) \propto e^{+i3\phi} + e^{+i\phi} - e^{-i\phi} - e^{-i3\phi}.$$

Comparison with Eq. (8.27) shows that a measurement of L_z can yield four possible values $+3\hbar$, $+\hbar$, $-\hbar$ and $-3\hbar$ with equal probabilities of 1/4.

8. (a) The eigenfunction with energy $\frac{3}{2}\hbar\omega$ has $l = 0$ and $m_l = 0$ because it is spherically symmetric.

(b) Form linear combinations of eigenfunctions with energy $\frac{5}{2}\hbar\omega$ which are propor-tional to $x + iy$, z and $x - iy$.

CHAPTER 9

1. (a) A spherical symmetric wave function implies zero orbital angular momentum.

(b) Evaluate the integrals

$$\langle V \rangle = \int_0^\infty N \mathrm{e}^{-ar} \left(\frac{-e^2}{4\pi\epsilon_0 r} \right) N \mathrm{e}^{-ar} 4\pi r^2 \, \mathrm{d}r$$

and

$$\langle T \rangle = \int_0^\infty N \mathrm{e}^{-ar} \left(\frac{-\hbar^2}{2m_e} \frac{\mathrm{d}^2}{\mathrm{d}r^2} r N \mathrm{e}^{-ar} \right) 4\pi r^2 \, \mathrm{d}r.$$

(c) Find the minimum of $\langle E \rangle = \langle T \rangle + \langle V \rangle$ by setting $\mathrm{d}\langle E \rangle / \mathrm{d}a = 0$.

2. (a) The minimum of $V_e(r)$ is found using

$$\frac{\mathrm{d}V_e}{\mathrm{d}r} = \frac{L^2}{mr^3} - \frac{e^2}{4\pi\epsilon_0 r^2} = 0.$$

(b) The maximum and minimum distances r, which occur when $p_r = 0$, are given by

$$\frac{e^2}{8\pi\epsilon_0 a} = \frac{L^2}{2mr^2} - \frac{e^2}{4\pi\epsilon_0 r}.$$

3. Use the integral given in problem 1 to evaluate

$$N^2 \int_0^\infty r^{2l+2} \mathrm{e}^{-2r/(l+1)a_0} \, \mathrm{d}r.$$

(b) Find the maximum of $r^{2l+2} \mathrm{e}^{-2r/(l+1)a_0}$.

(c) Use the integral given in problem 1 to evaluate

$$\int_0^\infty u_{0,l}^*(r) r u_{0,l}(r) \, \mathrm{d}r \quad \text{and} \quad \int_0^\infty u_{0,l}^*(r) r^2 u_{0,l}(r) \, \mathrm{d}r.$$

(e) For $l \gg 1$, the orbital angular momentum L tends to $l\hbar$ and $r_{\text{most probable}}$ and $\langle r \rangle$ both tend to $l^2 a_0$ or $L^2 a_0 / \hbar^2$.

4. (a) The eigenfunction is normalized if $N_1 = 1/\sqrt{\pi a_0^3}$.

(b) Show that

$$\int_0^\infty \psi_2^*(r) \psi_1(r) r^2 \, \mathrm{d}r = 0$$

if $\lambda = -1/2a_0$.

8. Choose $\mathbf{k}$ to be along the z axis so that $\mathrm{e}^{\mathrm{i}\mathbf{k}\cdot\mathbf{r}} = \mathrm{e}^{\mathrm{i}kr\cos\theta}$ and write $\mathrm{d}^3\mathbf{r} = r^2 \, \mathrm{d}r \, \mathrm{d}(\cos\theta) \, \mathrm{d}\phi$. Integrate from $\phi = 0$ to 2π, from $\cos\theta = -1$ to $+1$ and from $r = 0$ to ∞.

To find the most probable momentum, locate the maximum of $4\pi p^2 |\tilde{\psi}(p)|^2$. To find the average momentum evaluate the sandwich integral

$$\langle p \rangle = \int_0^\infty \tilde{\psi}^*(p)p\tilde{\psi}(p)\, 4\pi p^2 \, dp.$$

9. The simplest way is to write out the integral using Cartesian coordinates.

11. See Section 9.5.

14. When an electron is in the Coulomb potential due to a point charge Ze, the size of a bound state is proportional to $1/Z$ and the binding energy is proportional to Z^2.

CHAPTER 10

1. Exchange symmetry is a constant of motion if the Hamiltonian operator is unchanged when the particles are exchanged.

3. (a) Show that

$$\int_{-\infty}^{+\infty} dx_p \int_{-\infty}^{+\infty} dx_q \, |\Psi(x_p, x_q, t)|^2 = 1.$$

(b) Show that

$$\Delta E^{(S)} = \Delta E^{(D)} + K \quad \text{and} \quad \Delta E^{(A)} = \Delta E^{(D)} - K,$$

where K is given by

$$K = \int_{-\infty}^{+\infty} dx_p \int_{-\infty}^{+\infty} dx_q \, \psi_n^*(x_q)\psi_{n'}^*(x_p)V(|x_p - x_q|)\psi_n(x_p)\psi_{n'}(x_q).$$

CHAPTER 11

1. (a) The mean charge radius is given by

$$\int_0^\infty r\rho(r)\, 4\pi r^2 \, dr \bigg/ \int_0^\infty \rho(r)\, 4\pi r^2 \, dr.$$

(b) The charge inside radius r is given by

$$q(r) = \int_0^r \rho(r')\, 4\pi r'^2 \, dr'.$$

3. The effective point charges seen by the valence electron of a sodium atom when it is in a 3s, 3p and 3d state are

$$Z^*(3s)e = 1.84e, \quad Z^*(3p)e = 1.18e \quad \text{and} \quad Z^*(3d)e = 1.003e.$$

Note the shielding by the inner electrons increases with l.

8. The energy of a quantum state with spatial extent R of an ion with $Z - 1$ electrons and a nucleus with charge Ze is

$$E = (Z - 1)\frac{\hbar^2}{2m_e R^2} - Z(Z - 1)\frac{e^2}{4\pi\epsilon_0 R} + \frac{(Z - 1)(Z - 2)}{2}\frac{e^2}{4\pi\epsilon_0 R_{ee}}.$$

Further reading

Below is a short list of books on quantum mechanics for further reading. They vary in approach, topics covered and level of treatment, but all should be accessible to a student of this book.

Feynman, R.P., Leighton, R.B. and Sands, M., 1965, *The Feynman lectures on physics*, vol. III: *Quantum mechanics*, Addison-Wesley, Reading, MA.

Gasiorowicz, S., 1995, *Quantum physics*, 2nd edn, Wiley, New York.

Griffiths, D.J., 1995, *Introduction to quantum mechanics*, Prentice-Hall, Englewood Cliffs, NJ.

Liboff, R.L., 2002, *Introduction to quantum mechanics*, 4th edn, Addison-Wesley, Reading, MA.

Mandl, F., 1992, *Quantum mechanics* (Manchester Physics Series), Wiley, Chichester, England.

Park, D., 1992, *Introduction to quantum theory*, 3rd edn, McGraw-Hill, New York.

Peebles, P.J.E., 1992, *Quantum mechanics*, Princeton University Press, Princeton, NJ.

Rae, A.I.M., 2002, *Quantum mechanics*, 4th edn, Adam Hilger, Institute of Physics, Bristol, England.

Townsend, J.S., 1992, *A modern approach to quantum mechanics*, McGraw-Hill, New York.

Index

References to sections are printed in *italics*

PIGS IN CLOVER

OR, HOW I ACCIDENTALLY FELL IN LOVE WITH THE GOOD LIFE

SIMON DAWSON

LARGE
PRINT

First published in Great Britain 2015
by
Orion
an imprint of the Orion Publishing Group Ltd.

First Isis Edition
published 2016
by arrangement with
the Orion Publishing Group Ltd.

The moral right of the author has been asserted

A catalogue record for this book is available
from the British Library.

ISBN 978–1–78541–144–1 (hb)
ISBN 978–1–78541–150–2 (pb)

*For Debbie, what an amazing ride life's
turned out to be!*

*And for friends who shared some of the twists
and turns, Deacon and Darcy, Kylie and the
beautiful Bobby (Roberta's Pride).
Rainbow Bridge, guys, Rainbow Bridge. . .*